Introduction to Agricultural Statistics

Introduction to Agricultural Statistics

Bob Davis

Africa • Australia • Canada • Denmark • Japan • Mexico • New Zealand • Philippines
Puerto Rico • Singapore • Spain • United Kingdom • United States

NOTICE TO THE READER

Publisher does not warrant or guarantee any of the products described herein or perform any independent analysis in connection with any of the product information contained herein. Publisher does not assume, and expressly disclaims, any obligation to obtain and include information other than that provided to it by the manufacturer.

The reader is expressly warned to consider and adopt all safety precautions that might be indicated by the activities herein and to avoid all potential hazards. By following the instructions contained herein, the reader willingly assumes all risks in connection with such instructions.

The Publisher makes no representation or warranties of any kind, including but not limited to, the warranties of fitness for particular purpose or merchantability, nor are any such representations implied with respect to the material set forth herein, and the publisher takes no responsibility with respect to such material. The publisher shall not be liable for any special, consequential, or exemplary damages resulting, in whole or part, from the readers' use of, or reliance upon, this material.

Delmar Staff:
Business Unit Director: Susan Simpfenderfer
Executive Editor: Marlene McHugh Pratt
Developmental Editor: Andrea Edwards Myers
Executive Marketing Manager: Donna Lewis
Executive Production Manager: Wendy Troeger
Production Editor: Carolyn Miller
Cover Design: The Drawing Board

Printed in the United States of America
1 2 3 4 5 6 7 8 9 10 XXX 05 04 03 02 01 00 99

For more information, contact:
Delmar, 3 Columbia Circle, PO Box 15015, Albany, NY 12212-0515; or find us on the World Wide Web at http://www.delmar.com

Library of Congress Cataloging-in-Publication Data
Davis, Bob, 1938-
Introduction to agricultural statistics / Bob Davis.
p. cm.
ISBN 0-7668-1155-7
1. Agriculture—Statistical methods. 2. Agriculture—Statistics. I. Title.
S566.55 .D38 1999
630'.7'27—dc21

99-042442

Contents

Preface

This text is designed for a one-semester or term introduction to the statistical methods commonly taught to undergraduates in the agricultural sciences using examples and applications to which students can easily relate. It is presented in a logical sequence so that, as an instructor, you may find it natural to proceed from the first chapter to the last if time permits. If you are pressed for time, the material in the later chapters may be omitted without sacrificing the most important concepts. The exposition uses agricultural-based examples to teach the ideas and concepts of statistics. Statistical theory is woven into the narrative, but most of the concepts are presented using simple mathematics. Problems with an agricultural slant at the end of the chapters may be used to reinforce the ideas for students. Where practical, appendices to chapters are included that explain how to perform the statistical analysis on a personal computer spreadsheet program. Example problems that illustrate the use of the spreadsheet are included.

Chapter Descriptions

After an introduction to statistics and statistical inference in the first chapter, which ends with an appendix on summations, Chapter 2 covers the common measures of central tendency and dispersion used in summarizing both grouped and ungrouped data. This introductory material is then followed by chapters on probability and probability distributions, in which students are introduced to the main concepts of probability needed for working with probability distributions. Methods of calculating probabilities for the common distributions are presented with an emphasis on using the Appendix tables. Next, sampling and sampling distributions are introduced. If students can master the idea of a sampling distribution of the mean, for example, then the rest of the material on statistical inference is much easier to understand. Thus, in some ways, Chapters 3 through 6 represent key material for the course by providing a proper foundation for what follows.

The material on statistical inference begins with confidence intervals when there is one sample and then expands to two samples. Next, tests of hypotheses are discussed for the same cases as confidence intervals so that students may see their relationship. The material then covers the common one- and two-way analysis of variance models. In this section, the computer is especially useful to take the drudgery out of the calculations. Emphasis is placed on interpretation of the results. Test of hypotheses about proportions with three or more samples is next introduced using chi-square. The chi-square chapter also includes tests of independence and fitting distributions with chi-square.

Curve fitting is covered in Chapter 11, where simple correlation is introduced and simple linear regression is presented in detail. Multiple regression is introduced, and the chapter ends with time series, an important topic in agribusiness. Fitting linear and nonlinear trend with regression models is illustrated, and the chapter ends with a section on seasonal variation in time series data.

The last two chapters cover nonparametric statistical procedures including the Mann-Whitney U test, the sign test, the runs test, and rank correlation; and index numbers, another topic of importance to agribusiness students.

After taking a statistics course based on this text, students will hopefully not only understand the statistical methods, but will have an appreciation for the many types of problems in agriculture that can be analyzed statistically.

Acknowledgments

I wish to thank my wife, Anita, for her encouragement and support during this project and my colleagues at SWT, who were quite willing to discuss how we might use statistics in their fields of endeavor and who provided some much needed assistance in formulating problems for this text.

Bob Davis

The author and Delmar also would like to thank the reviewers who provided valuable insights on ways to improve the book; we sincerely appreciate their valuable time and advice.

Dr. John Foldy
The College of Saint Rose
Albany, New York

David Narrie
Tennessee Technological University
Cookeville, Tennessee

Joel Noel
California Polytechnic State University
San Luis Obispo, California

CHAPTER

1

Introduction

The purpose of this text is to provide an introduction to statistical inference in agriculture and the life sciences. Although we may not be statisticians when we have finished it, we will have gained an appreciation of the subject matter, learned how to use the most common statistical analyses, and acquired the ability to discuss complex problems with a professional.

What is statistics? Perhaps the most common conception of statistics is that of numerical data that serve as a record of the past. Statistics as numerical data can be dry bits of irrelevant trivia or interesting clues to future performance, depending upon their nature and their meaning to the person using them. For instance, during football season, what information is carried in the newspaper write-up of the game at the end of the column under the heading "statistics," and how useful are these "statistics"? It depends. If the coach of an opposing team reads them, then perhaps the number of yards passing, the number of interceptions, the average yards per punt, etc. are both interesting and useful. But a casually interested football fan might just as soon not see them since he might be satisfied with only the score.

So it is with agricultural data. Someone involved in the cattle-feeding business may actively seek information on the number of cattle on feed, the number of new placements last week, the average prices of feed ingredients, etc., so she or he can better manage the business. In this context, the principal factors that govern whether a set of statistics is useful are how knowledgeable a person is about the statistics being collected and how meaningful they may be with respect to future decisions. "Statistics," when used in this manner, is plural and always refers to numerical data. It is sometimes called **descriptive statistics.**

Statistics when singular, or statistical inference, refers to a process for making informed decisions. It involves inductive logic and the use of the scientific method. Its focus is upon problem solving. We recall that inductive logic involves reasoning from the specific to the general. In statistical inference, it is common to collect observations on a few items (a sample), analyze them using a well-known procedure, and then infer that since the findings obtain for the sample, they must obtain for all the items (the universe). This is the way the broadcast media forecast who has won the general election for President even before the polls have closed on the West Coast. They ask a few carefully chosen voters (a sample) how they voted and then infer that the choices made by the sampled voters will be repeated in the population of voters at large, i.e., they use inductive logic.

The scientific method describes a procedure used in the sciences to conduct experiments. It involves several steps: define the problem, formulate hypotheses, determine which data need to be collected and analyzed, select the form of analysis, and state how the outcome of the analysis bears upon the problem. See how statistical inference fits into this procedure? Once the problem and hypotheses are formulated, we make a decision concerning how to test—what data we need, how we will obtain, organize, and analyze them so that we will get an answer that bears on the problem—and statistical inference provides one method for doing this. We could also employ nonstatistical procedures, but we select the tools needed according to the nature of the problem, the type of analysis that best bears on the problem, and the kinds of conclusions we wish to make. Anytime we want to express the conclusions in the form of probability statements, or to be able to say how sure we are of the results in a probability sense, we must apply some form of statistical inference.

Perhaps now we can define statistics. Statistics involves the development and application of procedures and techniques for collecting, analyzing, and interpreting numerical data so that the reliability of the conclusions drawn from this process may be evaluated objectively with probability statements. The last part of the definition—the use of probability to evaluate the reliability of conclusions and inferences based on the data—is the unique and major contribution of statistics to inductive inference.

Common Misuses of Statistics

Disreali said that there were liars, damn liars, and statisticians, thus putting statisticians in with some very undesirable company. Huff has written a neat little book, *How to Lie with Statistics,* and Reichmann has also detailed the abuse of statistics. The fact that others have made similar comments would lead us to believe that not all who use statistical analyses do so with pure intentions. In business dealings, as well as in our personal lives, we are bombarded with presentations supported by statistics. So one reason to learn statistics is for self-protection! In the following section, we examine some of the more common ways that lies are told with statistics.

Bias

It is usually easy to detect conscious bias or favoritism in an advertisement that quotes statistics to prove the superiority of a given product, while a competitor's ad quotes other statistics to prove the opposite. However, unconscious bias creeps into statistical work because of biased data collected from questionnaires or from polls because of faulty sampling procedures, poor questionnaire construction, leading questions, or from an investigator who is trying to prove his or her own preconceptions. Anytime a study uses data, we expect the authors to state the source; if not stated, there could be a problem.

Faulty Generalizations

We often see faulty generalizations because the **sample** was too small, making it impossible to adequately represent the population to which the conclusions were applied. For example, it is difficult to represent fairly all the fish of a certain species in one of the Great Lakes with a sample of five. Generalizations from a sample of one are even worse, yet they are commonly made. Other faulty generalizations are made from nontypical samples, i.e., samples that are adequate in size but differ from the population in some important respect. For example, a survey of college alumni from the class of 1940 stated that their average salary was $95,107. The results were based on twenty-one returns from a survey of 142 class members. The results are faulty if those who returned the survey had higher incomes than the others, or if they exaggerated their earnings, as is sometimes the case. Also, if a few alumni who returned the survey had very high incomes, that greatly inflates the average.

Faulty Deduction

The common error of faulty deduction arises from our tendency to apply a valid general rule as if it were an invariant law. Thus, a government report's conclusion that deer grazing capacity is 10.2 percent greater than demand may be true for the nation, but not necessarily true for each region. There may be a shortage in the West and 15.5 percent excess capacity in the Southeast. If statements can be made as general tendencies that allow for individual differences, this problem is not as troublesome.

Noncomparable Data

Comparisons are frequently made between items that aren't really alike. For example, an automobile union representative stated that the price of a four-door family sedan has risen almost 50 percent in the last ten years. What he neglected to say is that during that period, the manufacturer added many new features to the car as standard equipment, thus making the new car different from the old. Errors due to noncomparability affect price indexes generally, since technology is constantly changing and component specifications vary from one time period to the next.

Errors in Semantics

Slanted or colored words are often used to influence the listener, as in political campaigns. Beware of leading questions in questionnaires that may suggest an answer. For example, the question, "*Do you like the cops*" will elicit a quite different answer from, "*What is your opinion of our police department?*"

Assuming Causation from Correlation

This statement means that because one thing precedes another in time, we assume the first is the cause of the second. For example, suppose you get caught in a thunderstorm, and then catch a cold. You assume that getting wet is the cause of the cold. This does not necessarily follow. In general, if factors A and B fluctuate together (A and B are correlated), it may be that A causes B, but it might also be true that B causes A, or that A and B influence each other continuously, or intermittently, or that A and B are caused by some other factor C, or that the correlation is due to chance. Thus, you may have been exposed unknowingly to a cold virus that attacked about the time you got wet. Statistics cannot be used to infer causation from correlation. If such inference is made, it must be based on the professional judgment of the investigator who is using what he or she knows about the factors A and B from his or her profession. All the statistics say is that the two factors move together in some fashion.

Oversimplification

A common error arises from oversimplifying a subject by omitting essential qualifications. The facts presented may be true, but will mislead the audience if other pertinent facts are omitted. For example, a recent tractor ad stated that sales were up 155 percent in Texas. Since when? By not providing information on the base year, the comparison is meaningless, yet it sounds great. Advertisers, politicos, and others are frequent users of this tactic.

Spurious Accuracy

Now that we live in the computer age, answers can be generated with a large number of trailing decimals. Unfortunately, the answers are no more accurate than the data. So if the data are accurate to the nearest $100, as might be the case for income data, we do not present the average income for the study as $44,203.50. While it may look impressive, the accuracy is spurious. Since the data in this case are accurate only to the last $100, we should present the number as $44,200.

The true meaning of facts is easily distorted. The statistical investigator must therefore be on guard to avoid misrepresenting the facts, and to detect misuse of statistics by others. A critical attitude is essential.

A Look Ahead

The material covered in the remainder of the text is divided into three parts: (1) statistics used in summarizing data so they can be more easily understood; (2) probability, probability distributions, sampling, and sampling distributions, which comprise the background needed for the material in the last section on (3) statistical inference.

We encounter descriptive statistics every day as we go about our routines. Someone is always presenting us with an average something or the other—maybe the average crop yield or the average moisture content of the last load of grain taken to the elevator, etc. Thus, we need to spend some time getting acquainted with these descriptive measures so we readily understand what they mean when someone presents them to us and when we want to use them ourselves as the one or two numbers that describe an entire data set.

Probability, probability distributions, etc. are needed background information for understanding statistical inference since almost all statistical analysis is based on probability. The glossary in the back of the book will help with unfamiliar terms as we go through this material. In this section, our focus is on basic concepts, which we will apply in the last part on estimation and tests of hypotheses.

The material on statistical inference presents ways to analyze data so we can determine whether one treatment really is different from another or not. And we can attach a probability statement to their difference, if the statistical test says they are different. While in some cases we can look at data and see that they are the same or different, in other cases that is just not the case. Statistical procedures have been developed over the last half century or so that take the guesswork out of trying to decide when a difference exists. We will spend considerable time learning these procedures. There are a number of common procedures, each developed for a specific set of problems, and we will examine them all.

References

Huff, D. *How to Lie with Statistics.* New York: W. W. Norton, 1954.
Reichmann, W. J. *Use and Abuse of Statistics.* London: Methuen Ltd., 1961.

Appendix: Variables and Summations

Variables

A **variable** has been defined as any object that can take on a range of values. In statistics, variables are usually denoted by common letters of the alphabet such as X and Y. This is purely convention or habit. We could just as easily give them names or use Greek letters, but we have to use some type of designation.

Since variables take on a range of values, they are defined by the types of values they assume.

Qualitative variables are not the result of measurement. Rather, they result from some attribute that cannot be measured numerically. Examples include marital status, good or defective parts, kinds of fruit, age groups, etc. These types of variables usually have values arbitrarily assigned to them, such as, let $X = 1$ if the ball is black, zero otherwise, or let $Y = 1$ for female, 2 for male, etc.

Quantitative variables refer to an object that can be measured or counted. If the variable can only be counted, it is discrete. **Discrete variables** take on whole number values. There are many such variables in statistics, such as the number of male pigs in a litter of eight, the number of arrivals at a co-op supply store in a given period of time, or the number of successes in n trials of an experiment. **Continuous variables** are the result of measurement; hence they can assume any value in the range of the variable, depending upon the accuracy of the measuring instrument. For example, if X is the output of fertilizer in tons per day from a fertilizer plant, then X can vary from zero to the maximum capacity of the plant, say 100 tons. Output is continuous because the units of measurement can be as fine as we want them: 100-ton units, tons, tenths of tons, hundredths of tons, thousandths of tons, etc., particularly if we have an electronic scale.

Summations

We use formulas in statistics as a general method of indicating how to compute a variable of interest. For example, the formula to compute the arithmetic average or mean of a data set tells how to make the computation regardless of the size of the data set we have selected. Thus we do not have to modify the formula when the size of the data set changes or the type of data in use changes. Formulas may contain summation signs, which help state how to perform the calculations. Therefore, we review the algebra of summations next.

Index of Summation

The summation operator Σ means to add up the values of the variable beginning with the first value stated in the index, progressing one value at a time and ending with the last value in the index. The first value of the index is written under Σ and the last value is written on top of Σ. For example, we evaluate the term in equation 1.1 by adding the ten values of X_i beginning with

$$\sum_{1}^{10} X_i \tag{1.1}$$

X_1, progressing through the index one unit at a time to X_2, then to X_3, and so on, ending with X_{10}. Since there are ten values in the index of summation, we add the first ten values of X. It depends upon the data set that X is a member of as to whether that includes all its values, but the formula says to add only the first ten. To add all the values of X, we rewrite the formula so that the letter n (generally) replaces the last number in the index, 10. Also, if the summation sign appears without an index, by convention the index is understood to go from 1 to n, and we sum all the values of the variable.

Summation of a Constant

If the letter k represents a constant, then the summation of k is evaluated as the number of terms in the summation multiplied by k. So we value the term in equation 1.2 by multiplying k by

$$\sum_{1}^{5} k \qquad \boxed{1.2}$$

5, the number of terms in the summation, and get the answer $5k$.

Summation of a Constant Times a Variable

If we sum an expression such as $2X_i$, where 2 is a constant and X_i is a variable, written $\Sigma 2X_i$, then we may factor the constant outside of the summation and write it as $2\Sigma X_i$. This last expression requires us to sum the variable and multiply the total by the constant 2.

An Algebra of Summation

The summation sign distributes over an expression in parentheses. Thus, we can rewrite the term $\Sigma(3X_i + 7)$ as $\Sigma 3X_i + \Sigma 7$, which we can further refine to $3\Sigma X_i + 7n$, where n represents the number of terms in the index of summation. We write the summation of a squared variable as ΣX_i^2 and call it a sum of squares. We evaluate it by first squaring each value of X and then summing, as in equation 1.3. On the other hand, we write a squared total as $(\Sigma X_i)^2$ and value

$$\Sigma X_i^2 = X_1^2 + X_2^2 + \ldots + X_n^2. \qquad \boxed{1.3}$$

it by obtaining the sum of X and squaring it. A sum of squares and a squared total are two very different concepts, and both are used frequently in formulas.

Exercises

1. Write out in full the sums represented by each of the following expressions:

   a. ΣX_i
   b. ΣX_i^2
   c. $\Sigma(Y_i + 4)$
   d. $\Sigma c X_i$
   e. $\Sigma f_i X_i^2$
   f. $(\Sigma X_i)^2$

2. Write each of the following expressions using a summation sign and appropriate limits:

   a. $X_1 + X_2 + X_3 + X_4 + X_5 + X_6$
   b. $X_1^2 + X_2^2 + X_3^2 + X_4^2 + X_5^2$
   c. $[(X_3 + 3) + (X_4 + 3) + (X_5 + 3)]^2$
   d. $(X_1 - a)^2 + (X_2 - a)^2 + (X_3 - a)^2$

3. If $X_1 = 2, X_2 = 7, X_3 = -8, X_4 = -1, X_5 = 6, X_6 = 3$, find the numerical values of a., b., and c. under question 2.

CHAPTER

2

Summarizing Data

When we obtain observations on one or more variables selected from a population or a sample, we have data. We learned that there are quantitative and qualitative variables, and these give rise to different forms of data. Quantitative variables are the result of measurement or counting. Thus, the numbers collected as observations on those variables are classified as ratio and interval data, respectively, since measurements give rise to ratios, and counts to intervals. On the other hand, qualitative variables that express attributes about a sample or population such as heifers, steers, bulls, and cows give rise to nominal data since these are names applied to categories the investigator is interested in as part of his or her analysis, or to ranks such as prime, choice, and select grades of beef, which order the carcasses from best to worst and hence produce ordinal data. In addition, we often categorize observations as to their source. If they are collected by means of an experiment or sample survey or from some published source that actually obtained the data, such as the census of agriculture, they are called primary data. However, if they are acquired from a source that did not collect the data even though the source may have published them, they are designated secondary data.

The purpose of data collection is to allow us to make informed decisions about the problem at hand. Through the manipulation and statistical analysis of the data, we obtain information that bears on our problem and use it to make a better decision, hopefully, than we would have made in the absence of that information.

We generally collect raw statistical data in random order, as shown in table 2.1. Since there is no real order to the data, we experience difficulty in obtaining anything of value from them upon inspection. We may **array** the data as an aid to interpretation (table 2.2). An array reorders the data from the smallest to the largest value. From an array, we readily compute the range by subtracting the smallest from the largest observation. For the cotton example, the range is 373 minus 215, or 158 pounds of lint. An array also indicates something about the distribution of

TABLE 2.1 Cotton Lint Yields in Pounds Per Acre from Seventy-Five Farms in the Blackland Prairie

255	373	242	257	305
285	358	279	261	312
288	297	303	290	283
215	299	275	260	284
311	217	295	260	316
314	290	294	280	326
249	256	276	292	286
333	274	250	295	318
292	283	274	336	325
290	309	272	296	346
254	235	268	299	352
334	251	367	309	315
228	259	354	283	306
234	258	365	281	312
342	268	278	291	288

TABLE 2.2 Array of Cotton Lint Yields in Pounds Per Acre from Seventy-Five Farms in the Blackland Prairie

215	217	228	234	235
242	249	250	251	254
255	256	257	258	259
260	260	261	268	268
272	274	274	275	276
278	279	280	281	283
283	283	284	285	286
288	288	290	290	290
291	292	292	294	295
295	296	297	299	299
303	305	306	309	309
311	312	312	314	315
316	318	325	326	333
334	336	342	346	352
354	358	365	367	373

TABLE 2.3 Frequency Distribution of Cotton Yields in Pounds of Lint Per Acre for Seventy-Five Farms in the Blackland Prairie

Cotton Yield	Number of Farms
215 up to 235	4
235 up to 255	6
255 up to 275	13
275 up to 295	21
295 up to 315	15
315 up to 335	7
335 up to 355	5
355 up to 375	4
Total	75

the units between the two extremes and their tendency to cluster toward some central value, such as 290 in the cotton example.

We may further summarize the data into a **frequency distribution** like that shown in table 2.3. We construct the frequency distribution with a given number of classes. Each class has a certain width or number of units expressed as an interval. For the data in table 2.3, there are eight classes, each with a width of 20 pounds of lint. The frequency column in the table expresses the number of observations that fall within each class. The total number of frequencies must equal the number of data points, or 75 in this example. The variable "per acre yield of cotton lint" is continuous. Thus we arbitrarily break the class intervals at 20 pounds for ease of presentation. We see that the cotton yield intervals in table 2.3 have the upper limit of the first class repeated as the lower limit of the second class. This signifies that we are using **real class** limits. The values in the first class can approach 235 but cannot equal 235. A data point with the value 235 is a member of the second class, which has 235 as its lower limit. When we use **stated classes,** they are constructed so that there is a break of one unit between the upper limit of the first class and the lower limit of the second, e.g., 215 − 234, 235 − 254, etc. We sometimes use stated classes in presentations because they cause less confusion in interpretation. However, we can use only real classes for computations such as determining the width of the class interval. If we use stated classes, we underestimate the interval width. We get the width of the class interval, 20, by subtracting the real class lower limit from the upper limit, e.g., 235 minus 215.

Not all frequency distributions have equal width classes. This is especially true when we are dealing with income data in which presentation with equal width classes requires that some classes be empty or contain few observations. We prefer to make the classes of different widths in these instances to both shorten the

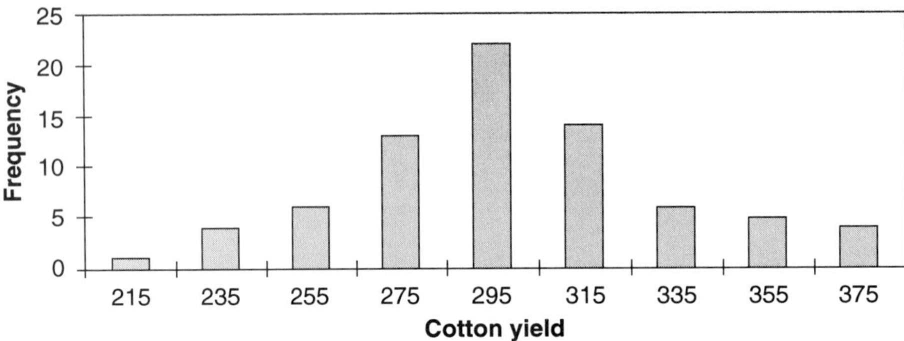

FIGURE 2.1 Histogram of cotton yields in pounds of lint per acre for seventy-five Blackland Prairie farms

frequency distribution and to aid in interpretation. With some data sets, we prefer to have the last class open-ended rather than have an upper limit because there are a few data points that are much larger than the planned upper limit of the class. In these cases, we write the last class as *355 and over,* or *more than 355.* A disadvantage of having open-ended classes is that we cannot compute some types of averages and measures of dispersion such as the arithmetic mean, midrange and standard deviation, and range from the frequency distribution.

A frequency distribution can also be presented in a bar chart, as shown in figure 2.1. This type of chart is called a **histogram.** In the histogram, the bars touch and the real class limits appear on the *x* axis at the end of each bar. If the class limits are not all the same width, neither are the bars of the histogram. The number of frequencies are displayed on the *y* axis to complete the chart.

An advantage of charting a frequency distribution is that we readily see its shape. It is easy to determine its skewness or if it is relatively symmetric in shape and to see the peakedness or kurtosis of the distribution. The cotton data in figure 2.1 are almost symmetrical. Also, the distribution is somewhat peaked or mesokurtic. It is not flat across the top or platykurtic.

The **frequency polygon** is a line graph that is sometimes used to display data. It shows frequencies on the *y* axis and class midpoints on the *x* axis. To make the graph touch the *x* axis, we plot midpoints of imaginary classes at each end of the distribution against zero frequencies. If we leave these out, the graph "floats" above the axis and looks strange. We compute class midpoints by averaging the upper and lower real class limits. For example, for the class *215 up to 235* of cotton data, we calculate the class midpoint as $(215 + 235)/2 = 225$, and so on for each class (see table 2.4).

Frequency polygons are even easier to use than histograms to tell the shape of a distribution. We commonly use frequency polygons to plot the relative frequencies on the same graph to compare two distributions, especially when we have a different number of frequencies in each distribution. We determine rela-

TABLE 2.4 Frequency Distribution with Class Midpoints for Cotton Yield Data, Blackland Prairie Farms

Cotton Yield	Class Midpoints	Number of Farms
Imaginary class	205	0
215 up to 235	225	4
235 up to 255	245	6
255 up to 275	265	13
275 up to 295	285	21
295 up to 315	305	15
315 up to 335	325	7
335 up to 355	345	5
Imaginary class	365	0
Total		75

TABLE 2.5 Frequency Distributions of Weekly Wages for Farm Workers and Truck Drivers

Weekly Wages	Class Mid-Points	Frequencies: Farm Workers	Frequencies: Truck Drivers	Relative Frequencies: Farm Workers	Relative Frequencies: Truck Drivers
100 up to 110	105	3	0	10	0
110 up to 120	115	7	0	23	0
120 up to 130	125	13	20	43	10
130 up to 140	135	5	48	16	24
140 up to 150	145	2	72	8	36
150 up to 160	155	0	44	0	22
160 up to 170	165	0	16	0	8
Total		30	200	100	100

tive frequencies by dividing the frequency in each class by the total number of frequencies for the distribution and then multiplying by 100 to get percent. As an example of how to compare two distributions, we examine the weekly wage distributions of farm workers and truck drivers, as shown in table 2.5. In these two data sets, farm workers' wages are generally lower; but when we plot them as frequency polygons (figure 2.2), we have difficulty in comparing the distributions because of the relatively small number of farm workers. The last two

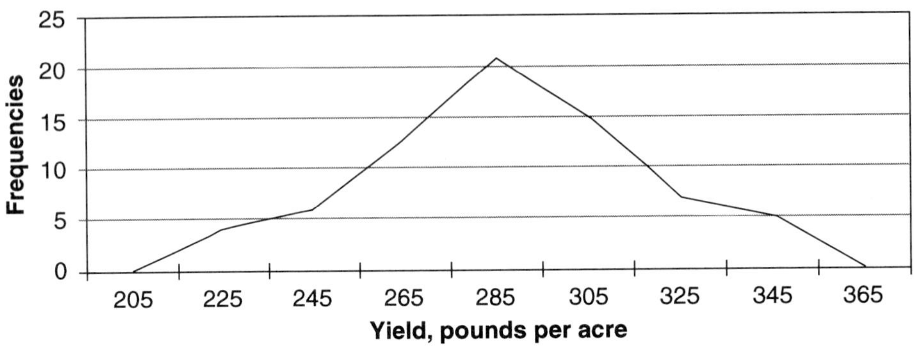

FIGURE 2.2 Frequency polygon of cotton yield lint data, Blackland Prairie farms

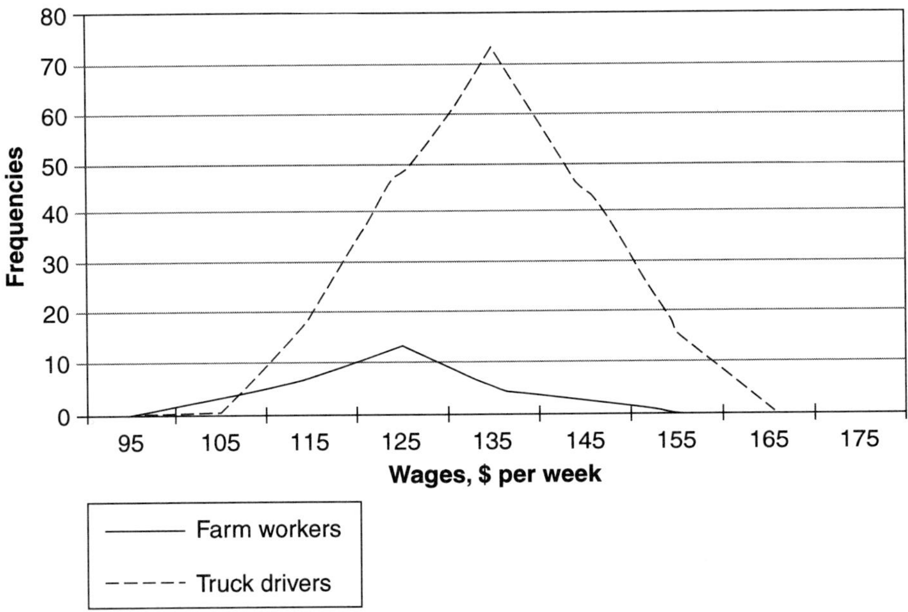

FIGURE 2.3 Frequency polygons of weekly wages for farm workers and truck drivers

columns of table 2.5 show relative frequencies for the two distributions, and relative frequency polygons for these data (figures 2.3 and 2.4), show that, in general, the wage distribution for truck drivers is higher than that for farm workers, but they have essentially the same shape.

We decide on the number of classes in a frequency distribution, but as a rule we select a value in the range of 5 to 15. Fewer than five classes does not provide enough categories to adequately display the data, and more than fifteen is gen-

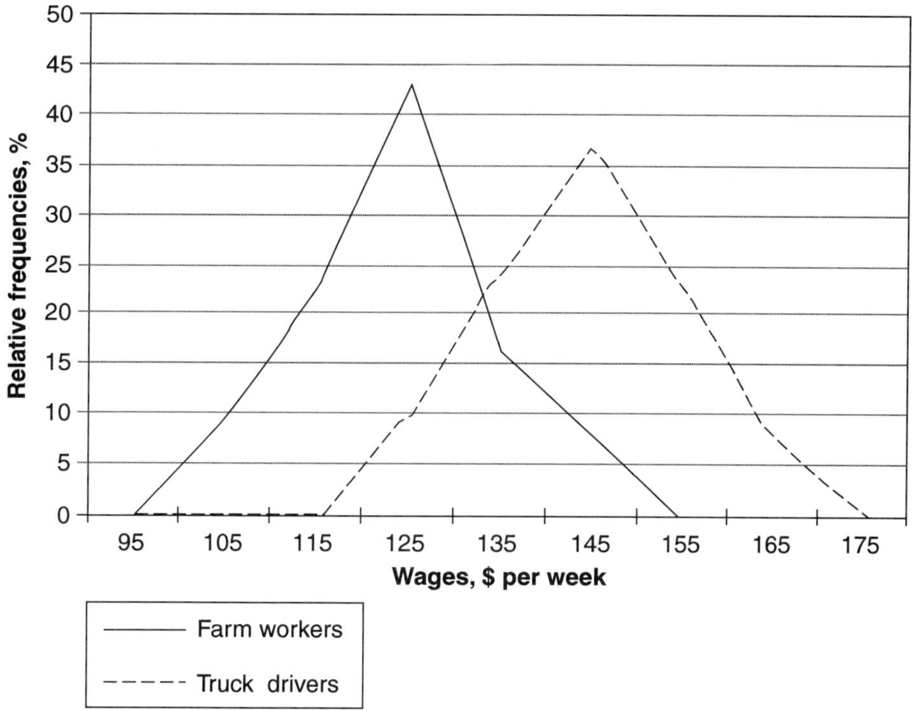

FIGURE 2.4 Relative frequency polygons of weekly wages for farm workers and truck drivers

erally too many. Sometimes we use **Sturges's rule,** which is a formula that solves for the number of classes, *K:*

$$K = 1 + 3.322(\log_{10} n)$$

where n represents the total number of frequencies.

The value K will usually not be a whole number, so we need to choose whether to round the number up or round down. Once we select K, then we obtain the interval width, I, by dividing the range of the data by K. When we determine values for the number of classes and the width of the class interval, we check the data to see that they work. Sometimes, one data point may be left out. Another consideration that is important in selecting interval widths is to make the class midpoints of the intervals representative of the data in the classes. For example, if we have data for wages that end in $5 values, we should select class midpoints that also end in $5 values. To do this, the class intervals may have to be adjusted. Sturges's formula is an approximation; we must finally decide what to do.

We can cumulate a frequency distribution to help understand its content. We cumulate either of two ways—on a "less-than" or "more-than" basis. In the

TABLE 2.6 Less-Than Cumulative Frequency Distribution of Weekly Wages for Farm Workers

Wages	Cumulative Frequencies, F
Less than 100	0
Less than 110	3
Less than 120	10
Less than 130	23
Less than 140	28
Less than 150	30

less-than procedure, we cumulate the frequencies beginning with the lowest class and ending with the highest. We write them as less than the upper limit of each class and add them from the top down (table 2.6). We can plot an S-shaped cumulative frequency curve called an **ogive** from these data with cumulative frequencies on the y axis and the upper class limits on the x axis (figure 2.5). Ogives provide an easy way to divide the frequency distribution into an equal number of parts such as percentiles, deciles, quartiles, and other quantiles. Percentiles divide data into 100 equal parts, deciles into ten equal parts, and quartiles into four equal parts. In the more-than procedure, we begin the cumulation with the highest class and end with the lowest. We state the values as higher than the lower class limit (table 2.7). We plot a more-than ogive with the cumu-

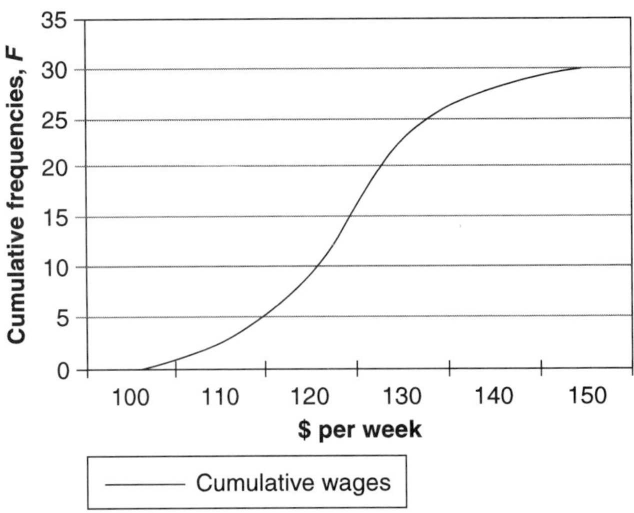

FIGURE 2.5 Less-than ogive of cumulative wages for farm workers

TABLE 2.7 More-Than Cumulative Frequency Distribution of Weekly Wages for Farm Workers

Wages	Cumulative Frequencies, F
More than 100	30
More than 110	27
More than 120	20
More than 130	7
More than 140	2
More than 150	0

lative frequencies on the y axis and the lower class limits on the x axis, and it makes a reverse S-shaped curve (figure 2.6). If we place the two ogives on the same graph, they cross at the middle of the distribution.

Averages

An **average** is a number used to represent the central value of a data set or a distribution. We use averages more often than any other statistical measure and express them both for ungrouped or raw unsummarized data and for data summarized into a frequency distribution. Computational procedures for the two

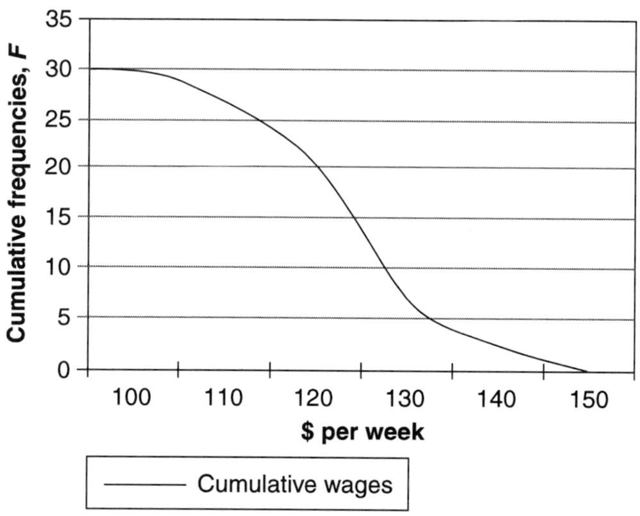

FIGURE 2.6 More-than ogive of cumulative wages for farm workers

types of data are generally different because interpolation formulas are often necessary when dealing with frequency distributions. Although the computation of averages from samples may be no different than that for the **population** or universe, we generally use different notations so there is no confusion as to which data we are manipulating. We recall that a sample is based on a subset of the data pertaining to the problem, while a population contains all of the data in the set. We use ordinary alphabetic characters in the formulas to express the average for sample data and generally employ Greek letters in formulas dealing with population data.

The Arithmetic Mean

Ungrouped Data

The **arithmetic mean** is the most widely used measure of central tendency. We obtain the mean, μ, of a population by summing the values of the observations, X_i, and dividing by the number, N, as in equation 2.1. For example, for the data set:

$$\begin{array}{r} 7 \\ 3 \\ 2 \\ \underline{8} \\ 20 \end{array}$$

$$\mu = \frac{\sum_{1}^{N} X_i}{N} \qquad \text{2.1}$$

the sum of the X_i is 20, the value of N is 4, thus, the mean,

$$\mu = 20/4 = 5.$$

If we take a sample from this population, say of size $n = 2$, we can calculate $\overline{X}$, the mean of the sample, by equation 2.2. The value of the sample mean depends upon which two

$$\overline{X} = \frac{\sum_{1}^{n} X_i}{n} \qquad \text{2.2}$$

items from the population we select. If we choose the first two values, 7 and 3, their sum is 10, and since $n = 2$, the sample mean is $10/2 = 5$, which is a perfect representation of the population mean. However, for several other choices of values from the population, the sample mean will not perfectly represent μ. It will be either too small or too large. Thus, the value of the sample mean depends upon which elements of the population we select for the sample, and it may or may not equal μ. The larger n becomes, the less effect extreme values in the pop-

TABLE 2.8 Hourly Wages and Number of Migrant Workers by Type of Crop

Crop	Hourly Wage, *X*	Number of Workers, *w*	*wX*
Cucumbers	4.50	950	4,275
Melons	4.75	600	2,850
Onions	5.25	1,020	5,355
Totals	—	2,570	12,480

ulation have on the value of the sample mean since they are combined with several other values, and the more closely the sample mean approaches μ.

In the preceding example, we gave each item from the population selected for the sample the same weight, namely a weight of 1. However, sometimes we want to assign weights other than 1, w_i, to the values of a set of data. In this case, we calculate the **weighted mean** by equation 2.3.

$$\overline{X} = \frac{\sum_{1}^{n} w_i X_i}{\sum_{1}^{n} w_i} \tag{2.3}$$

For example, the following data for the wages of migrant workers by type of crop also includes the number of workers for each crop. We can compute an average wage by using the number of workers as weights (table 2.8).

The weighted mean is computed as $12480/2570 = \$4.86$ per hour. It is appropriate because there is a different number of workers involved with each crop. An unweighted average would give a different result. In the case in which the weights are fractions and they sum to 1, the formula can be modified so that we drop division by Σw_i, since any value divided by 1 is unchanged.

Two properties of the arithmetic mean are: (1) the sum of the deviations from the mean is zero, if we define small x as a deviation from the mean as in equation 2.4, so that $\Sigma x = 0$;

$$x = (X - \overline{X}) \tag{2.4}$$

and (2) the sum of squares of the deviations from the mean is a minimum. If $M = \Sigma X/n$, then the quantity $\Sigma(X - M)^2$ is at least as small as when M is defined in any other way.

Grouped Data

When data are in a frequency distribution, often the values of the individual observations are not available and we must consider the distribution class by class. In this procedure, we use the class midpoint as a proxy for the values of all the

TABLE 2.9 Calculation of the Arithmetic Mean for Cotton Yield Data in a Frequency Distribution

Cotton Yield	Frequency, *f*	Class Midpoint, *X*	*fX*
215 up to 235	4	225	900
235 up to 255	6	245	1,470
255 up to 275	13	265	3,445
275 up to 295	21	285	5,985
295 up to 315	15	305	4,575
315 up to 335	7	325	2,275
335 up to 355	5	345	1,725
355 up to 375	4	365	1,460
Total	75	—	21,835

observations in that class. Also, since the frequency distribution has a different number of frequencies for each class, we use the frequencies as weights in calculating the arithmetic mean, as in equation 2.5. Suppose we decide to calculate the average yield of cotton lint from the frequency distribution in table 2.9. In this case, the arithmetic mean for the frequency distribution is $21835/75 = 291.1$, or 291 pounds of lint per acre, whereas the arithmetic mean for the raw data (see table 2.1) is $21807/75 = 290.8$, or 291 pounds. Thus the formula for the frequency distribution overestimated the mean by

$$291.1 - 290.8 = 0.3 \text{ pounds.}$$

$$\bar{X} = \frac{\sum_{1}^{n} f_i X_i}{\sum_{1}^{n} f_i} \tag{2.5}$$

Using the frequency distribution formula generally does not cost much in terms of lost accuracy provided we construct the distribution so that the class midpoints are good proxies for the data in the classes. This is good news because many secondary data sources, for example, most census reports, publish tables containing frequency distributions but generally do not present the raw data. Thus, if we wish to use frequency distribution data as the basis for calculating statistics such as the mean, we can do so with confidence.

The Midrange

The **midrange,** or center, is the arithmetic mean of the smallest and the largest items in a data set. When we array data, it is the average of the first and last items. Thus, in equation 2.6, if X_1 represents the smallest and X_n the largest, then we calculate the midrange, MR, as

$$MR = \frac{X_1 + X_n}{2}. \tag{2.6}$$

For the cotton yield data from table 2.2, the midrange is

$$(215 + 373)/2 = 294.$$

We use the midrange most often for computing the daily average temperature or average stock price because we are especially interested in the maximum and minimum values for these data series, and they are among the few values recorded. The midrange is generally unreliable when used as an estimate of the population mean because it is based on the two values in the data set that tend to change significantly from sample to sample, thus its value tends to vary widely in repeated sampling.

The Median

Unlike the arithmetic mean, which is based on the values of every observation in the data set, the **median** is a place average. It is not affected by the values of observations at the extremes of the data and is a better average to use in cases in which we encounter extreme values, as with income and education data. For ungrouped data, the median is the value of the middle observation after the data are arrayed. When we have an even number of observations, we average the middle two to obtain the median. For example, for the population of four values we had earlier:

7
3
2
8

we first array the data, and then select the middle observation:

2
3
7
8

We compute the median as the average of the two middle values, 3 and 7, which is equal to 5. The example we used in computing the median actually contains too few observations. Because the median is a position average, we should not use it with data sets that contain only a few observations. With six or fewer items in the data set, the median is probably a poor measure, but with twenty or more observations in the data set, it is likely to be very reliable. At least half of the observations in the data set are smaller in value than the median, and half are larger. For extremely peaked distributions, the median is the most reliable estimate of the population mean.

For grouped data, the median is determined by an interpolation formula, as in equation 2.7.

$$Md = L + \frac{\frac{n}{2} - F}{f} I \tag{2.7}$$

where: L is the lower limit of the class containing the median,
n is the total number of frequencies,
F is the cumulative frequencies in the class preceding the class containing the median,
f is the number of frequencies in the class containing the median, and
I is the width of the class containing the median.

To determine the class containing the median, we first cumulate the frequencies on a less-than basis. Since we know the median is the value of the middle observation in the data, we compute $n/2$ to determine the observation at the middle. We next inspect the cumulative frequency column and determine which class contains the middle observation. That class is the median class. Its lower limit is the value of L; its number of frequencies is the value of f; and its width is the value of I. To determine F, we go up one row and note the value of the cumulative frequencies for the class that precedes the median class. To determine the median, we plug all of the values into the formula and evaluate it.

For example, for the wage data for farm workers, the value of n, the total number of frequencies, is 30 and $n/2$ is 30/2, or 15 (table 2.10). To determine the class containing the median, we compare the value just computed, 15, to the cumulative frequency, F, column in the table. It falls between the values 10 and 23 for F. The value 10 signifies that the class *110 up to 120* contains seven values beginning with the fourth and ending with the tenth. Similarly, the value 23 means that the class *120 up to 130* contains thirteen values beginning with the eleventh and ending with the twenty-third observation. Thus, the fifteenth observation falls in the class *120 up to 130,* which is the class containing the median. So we write $L = 120, f = 13, I = 10$, and $F = 10$, which are the cumulative frequencies for the class *110 up to 120,* since that class precedes the median class. Plugging these values into the equation gives:

$$Md = 120 + \left(\frac{15 - 10}{13}\right)(10) = 120 + 3.8 = 123.8$$

The Mode

The **mode** is another average. It identifies the most common observation in the data set. For ungrouped data, we determine the mode by inspection. We simply examine the data and select the value of the observation that occurs most often. Ungrouped data may not have a mode if all values appear only once in the set; or we may observe several modes if there are many values that appear the same

TABLE 2.10 Frequency Distribution of Weekly Wages of Farm Workers Used to Calculate the Median

Weekly Wages	Frequencies, *f*	Cumulative Frequencies, *F*
100 up to 110	3	3
110 up to 120	7	*10*
120 up to 130	*13*	23
130 up to 140	5	28
140 up to 150	2	30
Total	30	

number of times in the data. For ungrouped data, the mode tends to be unreliable because its value changes significantly in repeated sampling. We use the mode when we want to know what is in vogue, or what is most common, such as the most popular television show, i.e., the one watched by the most viewers, or the most common variety of corn planted, etc.

For grouped data or distributions in general, the mode is the value at the highest point of the distribution. The crude mode is thus the midpoint of the class with the largest number of frequencies. For the data in table 2.10, the crude mode is 125 since the class *120 up to 130* contains the most frequencies, 13. We can use an interpolation formula for the mode (equation 2.8).

$$Mode = L + \frac{d_1}{d_1 + d_2} I \tag{2.8}$$

where: L is the lower limit of the modal class,
d_1 is the first difference: the frequency in the modal class minus the frequency in the preceding class,
d_2 is the second difference: the frequency in the modal class minus the frequency in the following class, and
I is the width of the modal class.

We can identify the modal class by inspection since it has the largest frequency. For the example in table 2.10, the modal class is *120 up to 130* because 13 is the largest frequency. Thus, L is 120; I is 10; d_1 is $13 - 7 = 6$; and d_2 is $13 - 5 = 8$. Plugging these values into equation 2.8 gives

$$\text{Mode} = 120 + \left(\frac{6}{6 + 8}\right)(10) = 120 + 4.3 = 124.3.$$

Thus, we get almost the same value as for the crude mode of 125.

When a distribution is bimodal, we often prefer to divide the data into two groups and analyze each group separately because some variable not accounted for in the analysis may be causing the data to cluster into the two groups.

Characteristics of the Mean, Median, and Mode

We may use the three averages together to determine the relative symmetry or skewness of a distribution. If the distribution is perfectly symmetric in shape, then the values of the three averages are identical. If the distribution has a tail on the right, or is skewed positively, then the arithmetic mean will be the largest, the mode the smallest, and the median about two-thirds of the way in between toward the mean. The mean is largest because it is affected by the few extremely large values in such a distribution. The median is sensitive to the position of the values, but not to their size. When a distribution has a tail on the left, or is skewed negatively, the mode is the largest, the arithmetic mean the smallest, with the median between the two about two-thirds the distance toward the mean. Thus, with any two of the three values, we have enough information to know whether a distribution is skewed negatively, is symmetric, or is skewed positively.

Occasionally we get a distribution with no mode, such as extremely platykurtic distributions as well as rectangular and uniform distributions.

The arithmetic mean is the only average of the three that we can use in algebraic calculations; hence for this reason alone, it is generally the most useful. For example, we can compute a grand mean from a set of sample means by averaging the means and using the number of observations in each sample as weights. However, the mean is difficult, if not impossible, to calculate when the frequency distribution contains open-ended classes. These do not affect the other two averages.

Measures of Dispersion

We define **dispersion** as the representativeness or validity of an average. It explains the amount by which data in a distribution, whether an array or a frequency distribution, are dispersed away from the average or clustered around it. The smaller the dispersion relative to the average it accompanies, the more representative the average, and thus the more comfortable we are in using the average as the one number to represent the entire distribution. For this reason, when we present averages for describing data, we should see that they are accompanied by some relevant measure of dispersion so the audience will know how well the average represents the data. For example, in a set of data on corn yields, if every yield is 100 bushels, the average is also 100 whether it is the mean, median, or the mode. The dispersion in this data is zero since the average precisely represents every item in the data set. But as soon as we observe only one yield value different from 100, we will see some dispersion, or scatter, in the distribution. The more different the individual corn yields become, the greater the scatter in the distribution and the less representative the average is as the one number to represent the set.

The Range

The **range,** R, is a measure of dispersion defined as the difference between the beginning and the end of a distribution when the data are arrayed, as in equation 2.9.

We can use it with the arithmetic mean, median, or the midrange. We do not use it with the mode. Indeed, the mode has no specific measure of dispersion associated with it, and partly for this reason, it is the least used of all averages. The particular advantage of the mode is for nonmathematical representations.

$$\text{The range } (R) = X_n - X_1. \qquad \boxed{2.9}$$

The range indicates not only how high and how low the values go, but the range of the data itself. We can use the range as a measure of dispersion for small samples ($n \leq 12$) selected from a normal distribution. However, for large samples ($n \geq 30$), the range provides a biased estimate of the variability of the population. The range is not our first choice as the measure of dispersion, because it is based only on the two extreme values of the data set—no others—and it tends to be inconsistent in repeated sampling from the same population.

The Quartile Deviation

The **quartile deviation** (QD) is a measure of dispersion used only with the median and indicates the dispersion of the data in the middle half of the distribution. We compute it as in equation 2.10.

$$QD = \frac{Q_3 - Q_1}{2} \qquad \boxed{2.10}$$

It is one-half the difference between the first and third quartiles. We remember that quartiles divide data into four equal parts, and to do this requires only three dividing lines: the first quartile, the median or second quartile, and the third quartile. Alternatively, we can view the first quartile as the median of the first half of the data, and the third quartile as the median of the last half. And for data in an array, that is how we compute the quartiles. First we determine the median for the data. We next examine the upper half of the data, from X_1 to Md, and determine the median of that set. We call it Q_1 since it is the first quartile. We employ the same procedure for the last half of the data to obtain Q_3. Finally, we compute the quartile deviation and use it with the median. For example, if we have the following years of formal education for a small sample of rural adults:

8
12
6
14
10

we first array the data to obtain:

6
8
10
12
14

The median of the data is 10; Q_1 is 7; Q_3 is 13; and thus

$$QD = (13 - 7)/2 = 3.$$

The quartile deviation is somewhat similar to the range because it includes measurement of the difference between two values, but the values are in the middle half of the distribution. Because it omits the ends of the data, the quartile deviation is a poor measure when there is wide dispersion in the tails of a distribution. It is quite useful, however, when we must compute an average for a frequency distribution with open-ended classes and we need a measure of dispersion to go with it. With frequency distributions, we use interpolation formulas for the first and third quartiles quite similar to that for the median, as in equations 2.11 and 2.12. We define all the terms in formulas 2.11 and 2.12

$$Q_1 = L + \frac{\frac{n}{4} - F}{f} I \qquad \text{2.11}$$

$$Q_3 = L + \frac{\frac{3n}{4} - F}{f} I \qquad \text{2.12}$$

the same as in the formula for the median except that we define the class containing the first quartile by comparing $n/4$ with the cumulative frequencies, F, and determine the class containing the third quartile by comparing $3n/4$ with the cumulative frequencies, F. Once we identify the appropriate class, L, f, and I refer to that class and F refers to the class that precedes it.

The Standard Deviation

We use the **standard deviation** as a measure of dispersion with the arithmetic mean. Its value is based on all of the observations in the data set. It is sometimes called the root-mean-square deviation because it is computed by taking the square root of the arithmetic mean of the squares of the deviations from the mean.

Ungrouped Data

We compute the standard deviation for a population, denoted σ (sigma), for which we have ungrouped data, as in equation 2.13. The population **variance** σ^2 is the square of the

$$\sigma = \sqrt{\frac{\Sigma(X - \mu)^2}{N}} \qquad \text{2.13}$$

standard deviation. We must use the variance in mathematical calculations, and then at the end, compute the standard deviation if that is what we want. The standard deviation is the most widely used of all measures of dispersion, in part be-

TABLE 2.11 Example Data for Calculating the Sample Standard Deviation

Days Absent, X	$(X - \bar{X})$	$(X - \bar{X})^2$
7	−3	9
14	4	16
8	−2	4
5	−5	25
15	5	25
11	1	1
60	0	80

cause the arithmetic mean is widely used as compared to other averages, and in part because the variance is mathematically sound and can be used in further calculations.

The sample variance, S^2, is an estimate of the population variance, σ^2, and as such is computed from sample data. We know that in repeated sampling, the sample variance is biased and underestimates the population variance by the fixed amount $(n - 1)/n$. Therefore, most statisticians revise the sample variance formula by dividing by $(n - 1)$ rather than n, which removes the bias. Thus, we have equation 2.14 for the sample standard deviation. We note that this revision is

$$S = \sqrt{\frac{\Sigma(X - \overline{X})^2}{n - 1}} \qquad \boxed{2.14}$$

significant when n is small, but not when n is large, say 100 or more.

To illustrate calculation of the sample standard deviation (table 2.11), we consider the following example: Suppose that in a large agribusiness firm, management selects the sick leave records of six employees at mid-year and records days absent from work as follows. Compute the arithmetic mean and sample standard deviation for these data. To calculate S, since the formula contains $\overline{X}$, we sum the first column of the table to get the data total, 60, and divide by the number of observations in the sample, 6, to get 10 for $\overline{X}$. We next compute the deviations from the mean by subtracting $\overline{X}$ from every value of X, one observation at a time, and recording them in the second column of the table. Finally, we square each value in column two of the table, recording the result in column three; then we total column three. S^2 is the value of $80/(n - 1)$ or $80/5 = 16$. So S is 4, the square root of the variance. The calculations in this example are relatively simple since the mean and the data values are all whole numbers. Thus, we get whole number values for the deviations from the mean in column two. When such values are not whole numbers, the calculations become somewhat tedious. However, we can use an alternative formula for S (equation 2.15) that does not contain deviations from the mean. To compute S by the alternate

$$S = \sqrt{\frac{\Sigma X^2 - \frac{(\Sigma X)^2}{n}}{n - 1}} \qquad \boxed{2.15}$$

formula, we need two columns for X and X^2, as follows:

	X	X^2
	7	49
	14	196
	8	64
	5	25
	15	225
	11	121
Totals	60	680

If we substitute 680 for ΣX^2, 60 for ΣX, and 6 for n in equation 2.15 and solve, we get $S = 4$, as we did before.

Grouped Data

For data grouped into frequency distributions, we use the frequencies as weights. Thus, we modify all of the formulas to incorporate the weights. The formulas for the standard deviation of sample data grouped into a frequency distribution appear in equations 2.16 and 2.17, with the definitional formula in equation 2.16 and the computational formula in equation 2.17. For an example for computing S, we once again refer to the data for wages of hired farm workers (table 2.12). Notice that the table, in addition to containing the original frequency distribution, has columns for X, fX, X^2, and fX^2, which we need to obtain values for the terms in our formula. We must

$$S = \sqrt{\frac{\Sigma f(X - \overline{X})^2}{\Sigma f - 1}} \qquad \boxed{2.16}$$

$$S = \sqrt{\frac{\Sigma fX^2 - \frac{(\Sigma fX)^2}{\Sigma f}}{\Sigma f - 1}} \qquad \boxed{2.17}$$

multiply f by X to get fX and then total those values to obtain ΣfX; we also square X and multiply f times X squared to get fX^2 and then total those values to obtain ΣfX^2. The only other value we need for the formula is Σf, which we obtain by totaling the frequency column. Thus

$$S = [(461950 - 3710^2/30)/(30 - 1)]^{0.5} = [108.5]^{0.5} = 10.4.$$

TABLE 2.12 Calculation of the Standard Deviation for a Frequency Distribution of Farm Wages

Weekly Wages	Frequencies, f	Class Midpoints, X	X^2	fX	fX^2
100 up to 110	3	105	11,025	315	33,075
110 up to 120	7	115	13,225	805	92,575
120 up to 130	13	125	15,625	1,625	203,125
130 up to 140	5	135	18,225	675	91,125
140 up to 150	2	145	21,025	290	42,050
Total	30	—	—	3,710	461,950

Two Properties of S and $\overline{X}$

We find two properties of the mean and standard deviation that are noteworthy, especially when we must make transformations to the data. For example, if we must add a constant to every element in the data set, what happens to the value of S? According to the first property, the value of the mean is increased by the value of the constant and S is unchanged by the addition of a constant to each value.[1] A second property is that multiplying each value of X by a constant multiplies the mean and S by the absolute value of the constant[2] and the variance by the square of the constant. We find these properties useful because we do not have to make the actual calculations for the mean or variance of the transformed data. We just apply the property to the mean or variance already calculated from the original data to obtain the value we want for the transformed data.

We can also standardize data by transforming the mean to a value of zero and the standard deviation to a value of 1. To obtain a mean of zero, we subtract the mean itself from every element in the data set one value at a time. The transformed data then have a mean of zero, but the standard deviation is not affected. To change the standard deviation to 1, we divide each element in the data by S (multiply by $1/S$). This calculation also divides the mean by S, but zero divided by any number except zero returns a value of zero. We generally denote the standardized data for a population by the variable Z, where $Z_i = (X_i - \mu)/\sigma$.

Coefficient of Variation

When we want a measure of the variability associated with a set of data, we use the standard deviation along with the mean. The standard deviation is in the same units as the mean and, hence, we prefer it to the variance, which is in squared units. But we often compute the **coefficient of variation** to determine the relative variability in a set of data. It is the ratio of the standard deviation to the mean, multiplied by 100 percent, i.e., $V = (S/\overline{X})(100)$. Thus, the coefficient of variation states how large the standard deviation is in comparison to the mean in percentage terms. A V of 100 percent indicates that S and the mean are equal. In that case, the data are highly variable and the mean is not a useful measure of the center of the distribution. The smaller the value of V, the better the mean represents the data set. As a rule of thumb, we use caution when we obtain values of V above 50 percent in deciding to represent the data set by the mean.

We cannot compare the standard deviations of two distributions because the S values belong to the means of those distributions, i.e., the standard deviation has no interpretation apart from the mean (except perhaps in the unusual case in which the means are equal). Thus, we compare the V values of the distributions instead. The distribution with the lowest V has the least variability.

Endnotes

[1]If $S^2 = \Sigma(X - \overline{X})^2/(n - 1)$ and we add a constant k to X so that we have $X + k$, then $\overline{X} = \Sigma(X + k)/n = \Sigma X/n + nk/n = \overline{X} + k$. Now $S^2 = \Sigma[(X + k) - (\overline{X} + k)]^2/(n - 1) = \Sigma[X + k - \overline{X} - k]^2/(n - 1) = \Sigma(X - \overline{X})^2/(n - 1)$, which is the original formula for the variance.

[2]If we have kX, then $\overline{X} = \Sigma kX/n = k\,\Sigma X/n = k\overline{X}$ and $S^2 = \Sigma[kX - k\overline{X}]^2/(n - 1) = \Sigma[k^2X^2 - 2kX\overline{X} + k^2\overline{X}^2]/(n - 1) = \Sigma[k^2(X^2 - 2X\overline{X} + \overline{X}^2)]/(n - 1) = k^2[\Sigma(X^2 - 2X\overline{X} + \overline{X}^2]/(n - 1) = k^2\Sigma(X - \overline{X})^2/(n - 1) = k^2S^2$ and, hence, kS for the standard deviation.

Exercises

1. Given the following sample of data, calculate: −10
  a. Arithmetic mean 2
  b. Median 3
  c. Mode 2
  d. Midrange −4
  e. Range 2
  f. Standard deviation 5

2. For the data in exercise 1, calculate the first and third quartiles and the quartile deviation.

3. Given the following set of numbers: 8, 5, 2, 6, 4, 5,
  a. Find the mean, median, and mode of this population.
  b. Find the variance and standard deviation.
  c. Calculate the range and midrange.

4. The following beginning salaries were offered to sixteen recent agriculture graduates:

$26,500	$19,900	$31,200	$31,400
$20,400	$21,400	$21,800	$25,500
$24,600	$22,600	$24,800	$27,000
$23,600	$28,400	$23,400	$29,100

  a. Compute the arithmetic mean and median for these data.
  b. Use a class interval of size $2,500 to form a frequency distribution for the data. Construct a histogram for this distribution and draw the frequency polygon connecting the class midpoints (begin with the class $19,000–$21,500).
  c. Form a less-than relative cumulative frequency distribution and plot the ogive. What percentage of the graduates earn less than $24,000? More than $29,000?
  d. Use the frequency distribution you obtained in part b to compute the arithmetic mean and standard deviation. What percentage of the salaries fall within one standard deviation of the mean? Two standard deviations?
  e. From the frequency distribution for part b., calculate the median and quartile deviation and the mode.

5. Given the following frequency distribution of the amount of calories in milk with 4 percent butterfat for a sample of 100 cows, compute the:

   a. Mean

   b. Standard deviation

   c. Median

   d. Mode

Calories	Frequency
90 < 100	12
100 < 110	55
110 < 120	25
120 < 130	8
Total	100

   e. Draw a histogram of the frequency distribution and comment on its symmetry.

6. How are the mean, median, and mode related in a symmetrical frequency distribution? How are they related in a frequency distribution that is skewed right? Skewed left? Sketch several distributions to clarify your answer.

Use Excel or a comparable computer program to complete the following exercise.

7. Given the following 205-day weaning weights of twenty-four black brangus calves, compute:

482	505	467	521
550	534	485	542
470	511	490	517
545	476	487	530
496	504	558	536
463	470	553	512

   a. The mean and standard deviation of the raw data using the Tools, Data Analysis, Descriptive Statistics module.

   b. A frequency distribution with intervals of 20 pounds beginning with 460 and ending with 560 using the Tools, Data Analysis, Histogram module. *Note:* In an adjacent column, construct the bin values 460, 480, 500, 520, 540, 560 prior to clicking Tools on the Menu Bar. Place this column range in the Bin Field on the Histogram definition table that appears. Also check the graph box so a graph will be generated.

CHAPTER

Probability

Statistical inference requires an understanding of the theory of probability to relate the properties of a sample to those of the population or universe from which it is drawn. Thus we need to know probability theory so that we can effectively consider statistical inference. Generally, we use probability theory to draw conclusions about the composition of a sample based on a mathematical model of the population. But we use statistical inference to make a conclusion about a mathematical model of the population on the basis of some statistic we computed from a sample.

The study of probability suggests three kinds of problems:

1. Defining and interpreting what is meant by probability,
2. Utilizing known probabilities to calculate others, and
3. Obtaining numerical probabilities.

We address the first two problems now, and consider the last later when we study the topic of estimation. Also, we discuss just the topics in probability theory that are most helpful in understanding statistical inference. We consider the meaning of probability and making deductions from the rules of probability. After we present some methods of counting, we examine probability theory, including topics from set theory, rules for calculating probabilities, revision of probabilities, and mathematical expectation.

Methods of Counting

We learn how to count early in life, and when we count, we arrange or list items. The type of counting that is important for probability theory involves choosing the number of ways we can arrange a set of items.

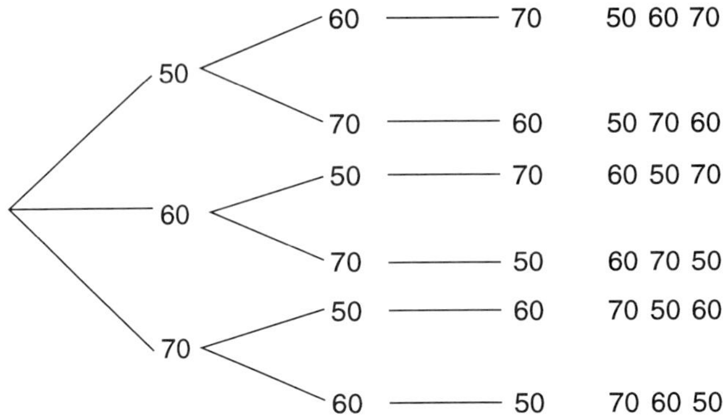

FIGURE 3.1 Tree diagram of possible permutations of three tractors

Permutations

A **permutation** is defined as any ordered sequence of a group or set of things. For example, assume a machinery dealer wishes to place three new tractors in a row in a showroom window. Further, assume the dealer wants to know the number of ways the tractors—Model 50, Model 60, and Model 70—can be arranged. One way for us to solve the problem is to list and count all possible arrangements of the tractors. The six possible arrangements, or permutations, are 50 60 70, 50 70 60, 60 50 70, 60 70 50, 70 50 60, and 70 60 50. We can use a tree diagram to present the arrangements graphically. The tree diagram (figure 3.1) shows every order in which we can place the tractors in the showroom, and we see that each change in order is a different arrangement.

Alternatively, we could consider the choices available at each step in selecting a tractor and arrive at the number of permutations that way. In the first step, there are three possible choices for selecting a tractor. But once we have selected the first tractor, there are two choices available for the second step. After we have selected two tractors, only one remains for the third step. Thus we have three ways to do the first step, two ways to perform the second step, and only one way to do the third step, for a product of (3)(2)(1), or six permutations or ways of selecting the tractors. We have just used the multiplication rule. As we might guess, many counting problems require more than three steps and can be quite involved in determining the number of permutations. So, let us state the multiplication principle more generally. If we can perform the first step in n_1 ways, the second step in n_2 ways, and so on for r steps, then the total number of ways we can perform the r steps is given by their product, as shown in equation 3.1.

$$(n_1)(n_2)(n_3)\ldots(n_r). \qquad \boxed{3.1}$$

For example, the luncheon menu of a restaurant offers a choice of three appetizers, five main dishes, four kinds of beverages, and six desserts. The total number of ways a different assortment of the luncheon items may be ordered is

$$(3)(5)(4)(6) = 360.$$

We can usually determine the number of permutations with the multiplication principle, but a shorter procedure works for particular kinds of problems. When we have every object in a set included in the permutation, we can calculate the number of permutations as $n!$, as in equation 3.2.

$$_nP_n = n! \qquad \textbf{3.2}$$

Consider the following example. Find the number of ways that four farm workers can be assigned to four different jobs. Using the $n!$ rule, we get

$$_4P_4 = 4! \text{ or } (4)(3)(2)(1) = 24$$

different ways of assigning all four workers to the four jobs.

For other counting problems, we are interested in the permutation of a subset r of the n objects in the set. In this case, we use the formula in equation 3.3.

$$_nP_r = n!/(n - r)! \qquad \textbf{3.3}$$

Consider the following example. Two salespeople are to be selected and given outside sales jobs from the six salespeople in the district office of an agricultural chemical company. The rest will remain inside. The number of possible assignments of the salespeople is

$$_6P_2 = 6!/(6 - 2)! = 6!/4! = 30.$$

In other kinds of counting problems, we want to know the permutations for each of two mutually exclusive activities, i.e., when one activity occurs, the other may not. In this case, if we perform the first activity in j ways and the second in k ways, then we can do one or the other of the activities in $j+ k$ ways. For example, if a tractor salesperson tells a farmer that the two-wheel-drive model is available with six power trains, two transmissions, and three cab types, while the front-wheel-drive-assist model is available in four power trains, three transmissions, and two cab types, then the customer has a total of

$$j + k = (6)(2)(3) + (4)(3)(2) = 60$$

different choices in selection of a tractor.

Combinations

While the permutations of a set of objects depends on the order of the arrangements, with each change in order reflecting a different arrangement, **combinations** do not depend on order. Thus, the number of possible combinations of all n objects in a set is one, since order does not count, and we have only one way

TABLE 3.1 Comparison of the Number of Combinations and Permutations of the Letters *wxyz* in groups of three

Combinations	Permutations
wxy	*wxy, wyx, xwy, xyw, ywx, yxw*
wxz	*wxz, wzx, xwz, xzw, zwx, zxw*
wyz	*wyz, wzy, ywz, yzw, zwy, zyw*
xyz	*xyz, xzy, yxz, yzx, zxy, zyx*

of including all of the objects in the combination. However, we may group r objects together from the n in a set and have more than one combination, which we denote by the symbol ${}_nC_r$.

In the following example, suppose there are four objects in a set, *wxyz*, and we want to determine the number of combinations and permutations of groups of three items at a time from the set. We can list them as shown in table 3.1.

Thus the number of combinations of four objects taken three at a time is four, the number of permutations is twenty-four. For each combination there are

$$r! = (3)(2)(1) = 6 \text{ permutations.}$$

When we divide the number of permutations, $n!/(n-r)!$, by the number of permutations per combination, $r!$, we obtain a formula for combinations, as shown in equation 3.4.

$${}_nC_r = n!/r!(n-r)! \qquad \boxed{3.4}$$

Consider the following example. A sales manager wants to place four different models of a tractor on a showroom floor. If six models are available, the number of combinations of four different models is

$${}_6C_4 = 6!/(4!)(6-4)! = 15.$$

We will see the same combinations formula with the binomial probability distribution later.

Probability

We can use **probability** whenever we need a conclusion concerning a matter that has an uncertain outcome. Probability theory offers us a way to mathematically estimate uncertain outcomes, and we want to choose between some deductions of probability and the mathematical interpretations we obtain from the postulates of probability.

However, we must define some terms before proceeding. The first of these is an **experiment.** An experiment is any process of observation or obtaining data. Thus, tossing dice, or observing the number of seeds that germinate in a flat of tomatoes, or recording the number of arrivals of customers at a supermarket check-out counter are experiments. Experiments have **outcomes.** The numbers that turn up on the dice, whether a particular seed germinated, and the number of arrivals at the supermarket check-out counter during the first hour on Saturday are all possible outcomes of an experiment. An elementary event, or an event, is the name we give to each outcome of an experiment that can occur on a single trial. For example, a trial of an experiment consists of tossing a die. The outcomes of the experiment are the numbers of dots 1 through 6 on the various faces of the die. Any number that actually turns up on a given toss of the die, or trial of the experiment, is an event. The numbers 1 through 6 are a complete list of the possible events, or an enumeration, of the possible outcomes of the experiment. Events are **mutually exclusive** if any one occurs and its occurrence precludes any other event. Consider a convenience store with two check-out counters. The check-out process at any one of the counters is an event. The events are mutually exclusive since a given customer checks out at only one counter.

Events have **observations,** or elementary units, associated with them. The sum of the elementary units comprises the population or the universe. Consider a bin that contains bolts of ¼″ diameter and other sizes. We could do an experiment by selecting a bolt at random from the bin, observing whether it is ¼″ in diameter, and replacing the bolt in the bin. The events are the two possible outcomes—¼″ in diameter or not ¼″. The universe of elementary units includes all of the bolts in the bin, and the two possible outcomes are associated with the elementary units.

We base objective probability interpretations on two different approaches to the meaning of probability and employ either theory or experience to determine probabilities. **Equiprobable events** suggest that if there is no reason to favor a particular outcome of an experiment, then we should consider all outcomes as equally likely. Thus we say that one side of a fair coin will turn up with a probability of ½ because there are only two possible outcomes—only one side can turn up, and we expect that both sides are equally likely to turn up when tossed. Hence the probability of ½ is the ratio of the number of ways in which a particular side can turn up divided by the total number of possible outcomes from the toss. Theory allows us to calculate the probability without finding a coin and actually tossing it. We call probabilities determined by using theory or intuitive judgment ***a priori*** probabilities.

We must have balanced probabilities such as drawing from a fair deck of cards or tossing a fair die before we can apply the idea of equiprobable outcomes. We do not need experimental data for such cases. However, even when we conduct an experiment under uniform conditions, the outcomes of the trials occur with differing frequencies. In other situations, we simply don't know the probability *a priori.* Thus we have devised a way to obtain probabilities under such conditions called **relative frequency.** The relative frequency, the number of times a certain event occurs in n trials of an experiment, provides us with a numerical estimate of the probability of an event. Unfortunately, the estimate depends upon the number of trials in our experiment. However, we find that when we conduct large numbers of trials in experiments in which we know the *a priori* probability, the relative frequency measure approaches the *a priori* value. This gives us the confidence to use the relative frequency measure of probability. Thus we define the probability of A, $P(A)$ by the expression in equation 3.5.

$$P(A) = \frac{\text{number of events favorable to } A}{\text{number of events in the experiment.}} \quad \boxed{3.5}$$

Basic Properties of Probability

Probabilities have three basic properties:

1. The ratio of the number of occurrences of an event A to the total number of trials must fall between 0 and 1, i.e., $0 \leq P(A) \leq 1$ if A is one of the mutually exclusive and exhaustive events of an experiment.
2. Since the events are collectively exhaustive, one of the elementary events must occur on a given trial. The probability that an event that occurs is *not* A is $P(A') = 1 - P(A)$. Thus, the sum of the probabilities of all the events must be 1.
3. If we examine the nature of A', we see that A' denotes an event composed of the mutually exclusive events other than A and we call it a **compound event.** Thus, the probability of A', $P(A')$ is the sum of the probabilities of all of the elementary events except A. For example, an ice cream manufacturer finds that on average, 60 percent of its customers purchases vanilla, 20 percent buys chocolate, 10 percent buys strawberry, and 10 percent buys all of the other flavors. The probability that a customer selected at random buys vanilla is $P(V) = 0.60$. The probability that a customer selects a flavor other than vanilla $P(V')$ is the sum of the probabilities for the remaining events, or $P(C) + P(S) + P(O) = 0.20 + 0.10 + 0.10 = 0.40$. Therefore, the probability of the compound event, not vanilla $P(V') = 0.40$ or $1 - P(V) = 1 - 0.60$.

An event with probability equal to zero means that it is highly unlikely to occur rather than impossible to occur. Likewise, $P(A) = 1$ does not mean that the event is certain to occur, but for all practical purposes, it will.

Relative frequency measures of probability have four basic features:

1. A large number of trials,
2. The relative frequency value approaches the *a priori* value, if available,
3. Use of empirical information gained from experience, and
4. Use of relative frequency to estimate probability.

For specific problems, the first two characteristics become limitations since, in some cases, we must determine probabilities before we can run trials. In such cases, we supply subjective estimates in place of actual or reliable experience. However, the relative frequency approach is widely used.

Sets

A **set** is a well-defined collection of things, either real or imaginary. By this definition, the employees of a grain elevator form a set. Likewise, a set might consist of the theories about the origin of the world, the farmers who carry crop insurance, or the outcomes of an experiment. The preceding are examples of discrete sets, which are composed of a finite number of elements. The following discussion concerns discrete sets.

We can identify a set in two ways:

1. We can list the elements of the set, or
2. We can determine whether a certain object belongs to the set by applying a rule.

When we list a set, the members are customarily enclosed within braces, {}. For example, to list the four suits in a deck of playing cards as the elements of a set gives

$$S = \{\text{clubs, diamonds, hearts, spades}\}.$$

The set of all the events that comprise an experiment is called the **universal set,** S.

Set Operations

In many cases, we are interested in only a portion (a subset) of the elements in the universal set. An **event** is any subset of a universal set of an experiment.[1] Events can have several elements, as in the following example. Suppose we toss two dice. Thirty-six possible outcomes can occur in this experiment; an outcome is the sum of the dots that turn up on any given toss. Suppose we are interested in the event, S_1, a subset in which the sum of the numbers that turn up is 7. We can list this subset as

$$S_1 = \{(1, 6), (2, 5), (3, 4), (4, 3), (5, 2), (6, 1)\}$$

if we have a way to tell the dice apart, such as using a different color for each.

We sometimes find it easier to determine subsets of the universal set if we can graph them. The Venn diagram is a type of graph in which we represent the universal set by a rectangle and the subsets as circles inside the rectangle. For example, we consider the set S of a deck of fifty-two playing cards with a subset of hearts, H, and a subset of kings, K.

We represent the set of all fifty-two cards by the rectangle shown in figure 3.2, and the subset of cards that are either hearts or kings by the areas enclosed by the circles H and K. The shaded and/or hatched area in the diagram is the subset of elements that are either in H or K, and it represents the union of H and K, written $H \cup K$, and read H or K, or H union K. The area where the circles

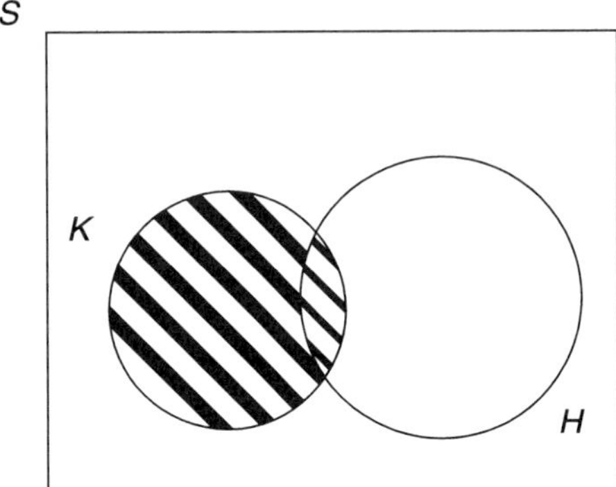

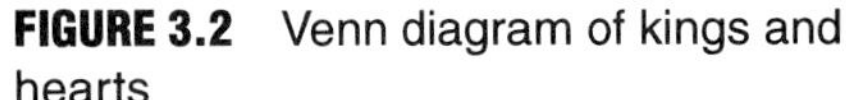
FIGURE 3.2 Venn diagram of kings and hearts

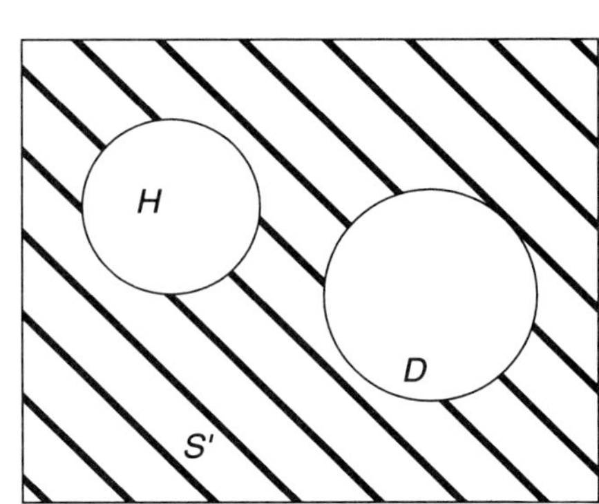

FIGURE 3.3 Venn diagram of disjoint and complement sets

overlap represents elements that are in both subsets, i.e., the king of hearts. We denote this portion as the **intersection** of H and K, written $H \cap K$, and read H and K, or H intersect K. We call subsets containing only one element like $H \cap K$ unit sets.

We can designate subsets from a universal set so they contain no common elements. When graphed, these subsets do not intersect, since if an element is a member of one subset, it cannot be a member of another. They are **disjoint.** To complete the designation of the areas of the diagram, the elements in the universal set S that are not included in one or more subsets of interest are denoted by S', which is called the **complement.** It means *not.* For example, we see the subsets of hearts and diamonds from a set of playing cards S in the Venn diagram in figure 3.3 as the areas H and D. The areas do not overlap since a card must be either a heart or a diamond or something else, but can't be both a heart and a diamond. The cards in the deck that are not hearts and diamonds, $S - (H \cup D)$, are represented by the striped area, S', in the rectangle.

This concludes our discussion of sets and set operations. We visit the ideas of unions, intersections, and complements later.

Random Variables

We recall that a variable is defined as any object that can take on a range of values. In the discussion on probability, we mentioned the idea of a variable when we were concerned with observing values such as those from the toss of two dice, or randomly selecting farmers for a survey. In these situations, we do not know the outcome of a trial in advance, but we have some information concerning the relative frequency of the outcome. We call systems such as these, in which the events are not identical or individually predictable with certainty, stochastic or chance procedures. Making repeated trials of an experiment involving stochastic processes yields results or events. A **random variable** takes on a

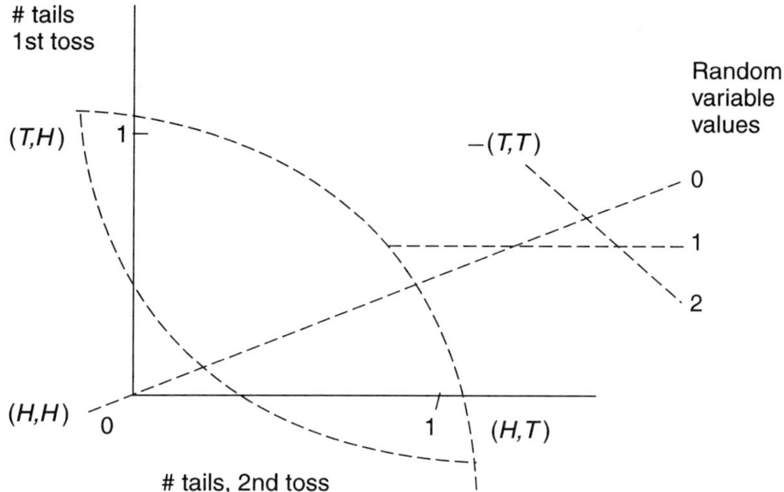

FIGURE 3.4 Graphical representation of sample space and random variable T for two tosses of a fair coin

definite value or property for each elementary event we are interested in (also called a chance variable or a stochastic variable). Random variables are of two types: continuous and discrete. We consider discrete random variables first.

The theory of sampling leads us to identify the elementary events or sample points from which we are collecting data. The **sample space** for an experiment is the set of all the sample points contained in it. Each point corresponds to only one possible event in the experiment. For example, let's assume we toss a fair coin twice in an experiment. On each toss, we get one of two possible outcomes, a head, H, or a tail, T. The sample space for this experiment is the set $\{(H, H), (H, T), (T, H), (T, T)\}$. The number of tails, T, that we can obtain is $\{0, 1, 2\}$, and T is a discrete random variable. The elements of the sample space identified with the various values of T are the subsets $\{(H, H)\}$, $\{(H, T), (T, H)\}$, and $\{(T, T)\}$. Thus, the possible number of tails in two tosses of the coin are contained in subsets of the events in the sample space. We show the sample space in this experiment graphically in figure 3.4 and display the values of the random variable in the column beside it. The lines in the diagram represent a function or counting rule. However, we are more interested in the function that links the values of the random variable to probabilities. We evaluate this function by comparing the number of points in our subset with those in the sample space. For our example, the proportion of outcomes in which one tail occurs, or $T = 1$, is $\frac{2}{4}$ or $\frac{1}{2}$; the fraction of outcomes giving $T = 0$, or no tails, is $\frac{1}{4}$, and the ratio for two tails is $\frac{1}{4}$. We now write the formula for this function in equation 3.6, which defines a binomial

$$P(r) = {}_nC_r\, p^r q^{n-r} \qquad \boxed{3.6}$$

probability distribution, which we will study in detail later.

We can express the concurrence between subsets of the sample space for an experiment and values of a random variable by selecting the proper equation. In fact, trying to determine which function to assign to a sample space is a significant problem in statistics.

Rules of Probability

The three rules of probability we introduced earlier help us in interpreting some measures of probability, but are not comprehensive enough for studying all of the probability problems that we might encounter in agriculture. So let's consider some additional rules of probability. The first three rules, by way of review are:

1. $0 \leq P(A) \leq 1$
2. $P(A_1) + P(A_2) + \ldots + P(A_n) = 1$
3. $P(A') = 1 - P(A)$

A fourth precept is

$$P(B) = P(A_1) + P(A_2).$$

This is the special rule of addition, which applies only to mutually exclusive events. If there are k mutually exclusive events, then we can expand the rule in equation 3.7 to make it more

$$P(A_1 \cup A_2 \cup \ldots \cup A_k) = P(A_1) + P(A_2) + \ldots + P(A_k). \qquad \boxed{3.7}$$

general. Consider, for example, a pen of 400 calves, of which 120 were sired by Bull A, 80 by Bull B, 96 by Bull C, and 104 by Bull D. If a calf is selected at random, the probability that it was sired by Bull A or B or D is

$P(A \cup B \cup D) = P(A) + P(B) + P(D) = 120/400 + 80/400 + 104/400 = 304/400 = 0.76$. By selecting a calf at random, we assure that each of the 400 calves in the pen has an equal chance of 1/400 of being chosen. If we are interested in selecting a calf sired by Bull A, we can obtain the probability as

$$P(A) = 120(1/400) = 120/400 = 0.30.$$

If the subsets corresponding to the events we are interested in have common elements, as in figure 3.2, then the events are not mutually exclusive. If the subsets intersect and are not disjoint, then we apply the general rule of addition (equation 3.8).

$$P(B_1 \cup B_2) = P(B_1) + P(B_2) - P(B_1 \cap B_2) \qquad \boxed{3.8}$$

In referring to figure 3.2, it is easy to see that we must subtract the intersection because we added it twice. The general rule of addition applies to probabilities whether or not they are mutually exclusive. When the events are mutually exclusive, the subsets are disjoint and the intersection is zero, which causes the general rule to be the same as the special rule of addition.

The rules of addition are useful for finding the probability of a compound event. However, in some probability problems we want to know the joint probability of several events occurring together. If the events are independent, their joint probability is defined by the special rule of multiplication, which we write as shown in equation 3.9 for k independent events $A_1, A_2, \ldots, A_k$.

$$P(A_1 \cap A_2 \cap \ldots \cap A_k) = P(A_1) \cdot P(A_2) \cdot \ldots \cdot P(A_k) \qquad \boxed{3.9}$$

Consider the following example. Records at a local university indicate that one-third of the agriculture majors are blond, one-half are junior college graduates, and three-fourths have attended classes there for more than one year. The probability that a randomly selected agriculture major is blond, a junior college graduate, and has been on campus for more than one year is, assuming independence,

$$(1/3)(1/2)(3/4) = 3/24 = 1/8.$$

Independent Events

Two events, A and B, are independent if the condition in equation 3.10 holds.

$$P(A \mid B) = P(A). \qquad \boxed{3.10}$$

This states that the chance of A, knowing B, is exactly the same as the chance of A without knowing B; or our knowledge of B does not change the probability of A. Therefore, A is statistically independent of B. On the other hand, if we have events A and C, and their relationship is as described in equation 3.11, then A is statistically dependent on C. Now our

$$P(A \mid C) \neq P(A) \qquad \boxed{3.11}$$

knowledge of C changes the probability of A. We call the expression $P(A \mid C)$ the conditional probability of A given C. So let's consider conditional probability in greater detail.

Conditional Probability

We frequently see problems that deal with only a portion of the sample space rather than all of it. Thus, the probability of an event depends upon what portion of the sample space we are considering. For example, the problem of selecting from all employees of a large agribusiness firm who have a college degree is different than selecting from the managerial employees of that firm who have a college degree. The management employees are a subpopulation defined by special conditions along with those attached to the total population. We define the probabilities associated with events in a subpopulation as **conditional probabilities.** For example, suppose the local auto dealer has sedans and sport utility vehicles (SUVs) with and without automatic transmissions for use as demonstrators, as described in table 3.2

TABLE 3.2 Number of Sedans and Sport Utility Vehicles (SUVs) Classified by Type of Transmission

	Sedans	SUVs	Total
Standard transmission	2	1	3
Automatic transmission	4	3	7
Total	6	4	10

If all of the cars are available, the sample space, S, is 10 for our random selection of a vehicle. The probability of an event A, the choice of an auto with an automatic transmission, is $P(A) = 0.7$; the probability of an event B, the random preference of a sport utility vehicle, is $P(B) = 0.4$. Now we assume a customer asks to drive a SUV. We reduce the sample space from $S = 10$ to the subpopulation or subspace, $B = 4$. The conditional probability of A (auto with an automatic transmission), given that it is a SUV, is $P(A \mid B) = 3/4$, which is the probability of A associated with the reduced sample space B. We can verify the result $P(A \mid B) = 3/4$ by considering that $P(A \mid B)$ is equal to the $P(A \cap B)$, $3/10$, the proportion of all autos that have automatic transmissions and are SUVs, divided by $P(B)$, $4/10$, the proportion of all autos that are SUVs, as given by equation 3.12.

$$P(A|B) = \frac{P(A \cap B)}{P(B)} \quad \text{3.12}$$

This definition of conditional probability requires that $P(B) \neq 0$.

The General Rule of Multiplication

If A and B are dependent events, we can express the probability that A and B both occur, or their joint probability, by either relationship in equation 3.13. Consider the following example.

$$P(A \cap B) = P(B) \cdot P(A \mid B) \text{ or}$$
$$P(A \cap B) = P(A) \cdot P(B \mid A). \quad \text{3.13}$$

Suppose a worker in a shop may select from a half dozen identical power drills. Unknown to the worker, four of the drills work and two do not. If the worker randomly selects two drills, what is the probability that both work? Since the worker is sampling without replacement, we have the probability that the first drill works, $P(W_1) = 4/6$, and the conditional probability that the second works, given that the first works, $P(W_2 \mid W_1) = 3/5$. Thus, the joint probability that both drills work is

$$P(W_1 \cap W_2) = P(W_1) \cdot P(W_2 \mid W_1) = (4/6)(3/5) = 12/30 \text{ or } 2/5.$$

Mathematical Expectation

The concept of mathematical expectation is one that we encounter frequently in later chapters. We can relate it to the ideas of probability we have just covered. The **expected value,** or **mathematical expectation,** of a variable is the probability-weighted arithmetic mean of the variable. That is, the weights are the probabilities associated with each of the values that the variable can assume. For example, we recall an experiment described earlier concerning the number of tails, T, obtained from two tosses of a fair coin. The random variable T can take on the values 0, 1, and 2 and has a probability corresponding to each value of T, *viz*, 1/4, 2/4, and 1/4. The expected value of T therefore is the sum of the product of each of its possible values and the respective probabilities, as given in equation 3.14. For our example, the expected value or average

$$E(T) = \Sigma T_i \cdot P(T_i)$$
$$E(T) = (0)(1/4) + (1)(2/4) + (2)(1/4) = 0 + 2/4 + 2/4 = 1. \quad \boxed{3.14}$$

value of T is one; or we expect one tail on average from two tosses of a fair coin. The expected value may or may not be a value that the random variable can assume. We interpret it as the long-run average for the variable for many repeated trials of the experiment; or as the probability-weighted mean of the values of the variable.

Consider another example. Suppose a player pays \$2 to play a game of chance as follows. He draws one card from a well-shuffled deck of playing cards. If he gets the queen of hearts, he wins \$15; if he draws any other heart, he wins \$1; and if he draws any other card, he loses his \$2. What is the expected value of this game? For the game to be fair, the expected value must be zero. If it is positive, the game favors the player; if negative, it favors the house. The expected value is obtained by placing the appropriate values in equation 3.14 and solving.

$$E(X) = \Sigma X \cdot P(X) = (\$13)(1/52) + (-\$1)(12/52) + (-\$2)(39/52) = -\$1.48.$$

Since the expected value is negative, the game is unfair. We interpret the expected value in this manner. In the long run, each time the player plays, he loses \$1.48. As we might guess, the idea of mathematical expectation was developed for games of chance and lotteries, but now it is a common term and we use it everywhere for evaluating alternative courses of action.

Endnote

[1] Notice that this is not the same as an elementary event as defined earlier in the context of probability analysis. Unfortunately, both set theory and probability theory contain events.

Exercises

1. Let $A = \{3, 4, 5\}$, $B = \{4, 5, 6, 7, 8, 9\}$, and $C = \{4, 6, 8, 10\}$.
   - **a.** List all the subsets of A. What is the complement of $\{4, 5\}$?
   - **b.** Find $A \cap B$ and $A \cup B$. Use a Venn diagram to depict these sets.
   - **c.** Find $A \cup B \cup C$, $A \cap B \cap C$, and $(A \cup B) \cap C$. Illustrate the sets on a Venn diagram.
2. Let $S = \{2, 4, 6, 8, 10\}$

   $A = \{2, 4, 6\}$
   $B = \{6, 8\}$
   $C = \{4, 6, 8, 10\}$

   Define the following sets:

   - **a.** $A \cap B$
   - **b.** $A \cup C$
   - **c.** $(A \cup C)' \cap B$
   - **d.** $A \cap C'$
3. If a single die is tossed 300 times, how many times would you expect a 3 to turn up? What is the probability of getting a number higher than a 3?
4. If two dice are tossed, what is the probability of getting:
   - **a.** A 7?
   - **b.** An even number?
   - **c.** A 7 or an even number?
   - **d.** A 7 or an 11?
   - **e.** Less than a 7?
   - **f.** The intersection of less than a 7 and an even number?
5. Suppose a box contains four chips numbered 1 through 4.
   - **a.** Enumerate the sample space for the experiment; draw two chips with replacement.
   - **b.** What is the probability that the two chips drawn in the experiment described in a. sum to less than 5?
   - **c.** What is the probability that the two chips drawn in the experiment described in a. sum to less than 5, given that the first chip was a 2?
6. You draw playing cards without replacement from a well-shuffled deck of fifty-two.
   - **a.** What is the probability that on two successive draws you get a king and a diamond ?
   - **b.** What is the probability that on five successive draws you get five diamonds?
   - **c.** In two successive draws, what is the probability of drawing a 10 on the second draw, given that a 10 or a jack was drawn on the first draw?
   - **d.** In a poker hand, what is the probability of receiving four aces and some other card?

7. **a.** In how many ways can four show calves be placed from first to fourth?
   b. In how many ways can six show calves be placed from first to fourth?
   c. How many different combinations are there of four calves in the first four placings from a group of six?

8. An agricultural equipment dealer has two round-bale and four square-bale balers. If the dealer wants to display all of them in a row on his or her lot, in how many different arrangements can they be displayed?

9. A student club has ten female members and twelve male members. If a homecoming float design committee of six students is to be randomly selected from the membership, what is the probability that four committee members will be female?

10. If male and female births are equally likely, what is the probability that only two males will be born in a litter of eight pigs? At most two males?

11. Graph the probability distribution for the random variable X, the number of heads in three tosses of a fair coin; also set up a table showing the values of X and the probability of each value. Compute the expected value of X. Draw a tree diagram representing all of the possible outcomes.

12. A student club decides to have a raffle to raise money. Five hundred tickets are printed with a face value of $2 to raffle off a $200 shotgun.
    a. What is the probability distribution for the random variable X, the amount of winnings for one randomly selected ticket buyer, if all tickets are sold?
    b. Compute the expected value of X.

13. The following table refers to carrots grown on experiment station plots, harvested, graded, and sorted by color.

	Grade			
Color	*Number 1*	*Number 2*	*Cull*	**Total**
Orange	900	400	200	1,500
Maroon	600	300	100	1,000
Total	1,500	700	300	2,500

Calculate the probability that a carrot randomly selected from this group will be:
a. An orange cull carrot.
b. A Number 2 grade.
c. A Number 1 grade, given that the carrot is maroon in color.
d. Either a Number 2 grade or a maroon carrot.
e. A maroon carrot, given that it is a Number 2 grade.

14. A roulette wheel has thirty-eight slots, 0, 00, and the numbers 1–36, in which the ball may land. The zeros are green, the even numbers are black, and the odd numbers are red. If the ball is thrown randomly in a slot, what is the probability that:

a. It lands on black?
b. It lands on the number 10?
c. It lands on either black or the number 10?
d. It lands on green or red?
e. If you play red an infinite number of times, you win what proportion of the time?

CHAPTER

Probability Distributions

Sometimes we are interested not only in the probability associated with a particular event, but also in the distribution of probabilities over the whole range of possible events. For example, let's consider a fair die with faces numbered 1, 2, . . . , 6. If X represents the random variable, the number of dots that turn up on the face of the die, we can write down the probabilities associated with each value of X as in equation 4.1. All of the probabilities are the same because we assume the

$$P(X = 1) = 1/6, P(X = 2) = 1/6, \ldots, P(X = 6) = 1/6 \quad \textbf{4.1}$$

die is fair. The listing of the possible values of X and their associated probabilities is called a **probability distribution.** We can avoid writing each separate probability by using a formula such as equation 4.2. We note that the probabilities summed over all the possible values of X add to 1.

$$P(X) = 1/6 \text{ for } (X = 1, 2, \ldots, 6). \quad \textbf{4.2}$$

This is a characteristic of all probability distributions and it means that all possibilities are accounted for. We remember that all probabilities must be positive, or non-negative.

Probability distributions are somewhat related to frequency distributions. To review, a frequency distribution is a listing of all of the possible outcomes of a variable that has been divided into classes, along with the frequency associated with each class. A probability distribution lists all of the possible outcomes of a random variable along with the probability associated with each outcome, rather than frequencies.

The purpose of many statistical problems is to select the best probability distribution to portray the random variable, as we mentioned before. Prior to examining the distributions used most often in statistics, we look at probability density functions.

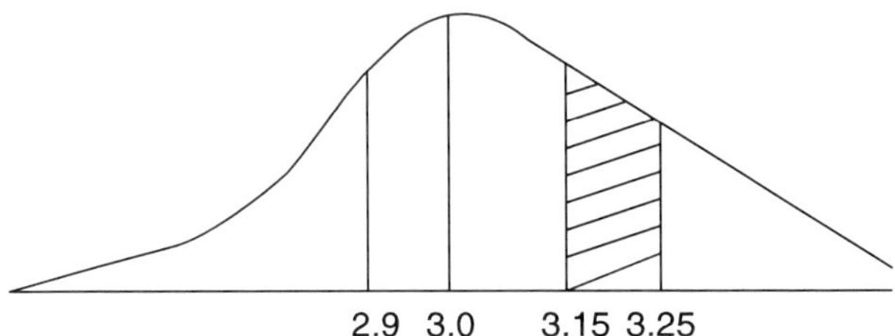

FIGURE 4.1 A continuous probability distribution

Probability Density Functions

A **probability density function,** abbreviated pdf, may be either discrete or continuous. A discrete pdf is a point function defined over a finite sample space; hence it takes on only a finite number of values. The pdf assigns a probability to each possible outcome of the discrete random variable. A continuous pdf is a set function that represents a distribution in which a probability is assigned to a range of values from a continuous random variable. The continuous pdf assigns a probability to the range 3.15 to 3.25 for the continuous random variable X, the diameters of shafts produced by a farm machinery company (figure 4.1).

A pdf is set apart from a probability distribution in that a pdf is really a rule for assigning a probability to the events of an experiment, while a probability distribution is an orderly presentation or arrangement of the probabilities once they have been assigned. With this in mind, let's examine a few of the more common probability distributions.

Binomial Probability Distribution

A frequently encountered set of statistical problems involves random events for which there are only two possible outcomes. The outcomes occur without any fixed pattern, and the probability of either outcome remains fixed for each trial. These problems are called **Bernoulli trials.** Bernoulli systems have two outcomes—the outcome we want or are interested in, called a success, and the other outcome, called a failure. We sometimes know with certainty the proportion of successes obtained from a very large number of trials and call it a parameter.

The probability density function that comes from Bernoulli trials is the binomial pdf, which gives the probability, p, of r successes in n trials of an experiment. Of course, if we know the number of successes, r, we also know the number of failures, $n - r$, since there are only two outcomes to the experiment. We can easily compute the probability of a failure as $1 - p$, which we sometimes write as q. The formula for the binomial pdf is presented in equation 4.3.

$$P(r \mid np) = {}_nC_r\, p^r (1 - p)^{n-r} \text{ or } {}_nC_r\, p^r q^{n-r} \tag{4.3}$$

Consider the following example. Suppose we toss a fair die four times and want to know the probability of obtaining three 1s in those four tosses. First, we compute the number of ways of obtaining three 1s in four tosses by using combinations; thus

$$_4C_3 = 4!/(4-3)!3! = 4 \text{ ways.}$$

We obtain the probability that any of the four ways will occur by the multiplication rule. For instance, the probability of obtaining the sequence of three 1s (1) and one value other than 1 (1') is

$$P(1111') = (1/6)(1/6)(1/6)(5/6) = 5/1296.$$

Thus, the probability of getting three 1s in four tosses of a die is

$$(4)(5/1296) = 20/1296 \text{ or } 5/324.$$

We can express the same problem using the binomial pdf as

$$_4C_3\ pppq \text{ or } {}_4C_3\ p^3q^1 \text{ or } (4)(1/6)^3(5/6)^1 = 5/324.$$

Let's consider another example and this time use equation 4.3 to solve it. Suppose a litter of ten pigs is born and we want to know the probability that it contains only one male. We can use the binomial pdf with $r = 1$, $n = 10$, and $p = 0.5$. Thus

$$P(r = 1 \mid n = 10, p = 0.5) = {}_{10}C_1\ (0.5)^1(0.5)^9 = 10!/1!9!(0.5)(0.00195) = 0.0098,$$

which makes it very unlikely.

The term *binomial* comes from the way the pdf is set up. Its value for given values of r, n, and p is equivalent to the corresponding term of the expansion for the binomial expression $(p + q)^n$.

The formula for the binomial pdf provides a means of computing the probability for every possible number of successes in a given number of trials. An orderly arrangement of these probabilities is called a binomial probability distribution. Thus, the binomial pdf defines an entire family of distributions, one for every combination of values for n and p, the parameters of these distributions. While only distributions with the value of $p = 0.5$ are symmetric, distributions with small values of p are reasonably symmetric, especially for large n. Thus, we can use the normal curve, which is symmetric, as an approximation to the binomial, and vice-versa.

There are cases when we want to find the probability of "r or more" successes or of "r or less" successes in n trials. In effect, what we want is the probability corresponding to the area under one of the tails of a binomial distribution. We obtain such probabilities by adding the binomial probabilities for the particular events. For experiments with large numbers of trials, this summation can be long and boring, especially if the value of r is not near zero or near n. For this reason, tables of binomial probabilities are published that give the tail areas for a large range of values for n, p, and r. See Appendix table 2, page 221 for an example table.

TABLE 4.1 Outcomes of Seven Trials in Which Male and Female Animals are Selected Randomly

Trial	Outcome	Value of X	Variable Defined
1	F	1	X_1
2	M	0	X_2
3	F	1	X_3
4	F	1	X_4
5	M	0	X_5
6	F	1	X_6
7	M	0	X_7

The Mean and Variance of a Bernoulli Process

It is fairly easy to derive the mean and variance of the binomial distribution because we can depict Bernoulli trials in several ways. Suppose we have an experiment in which two animals are drawn, with replacement, from a large pen in which half of the animals are male, M, and the other half are female, F. If the total number of animals is N, then the proportion of females is $F/N = p$. Thus, the probability of a success, selecting a female animal, F, is $p = 1/2$ and the probability of a failure, selecting a male animal, M, is $(1 - p) = 1/2$. Rather than run this through the binomial pdf, let's use a different approach. Assume we draw a random sample of n animals from the pen and we have the discrete random variable X that takes on the value 1 for a success (getting a female, F) and a value of 0 for a failure. In this manner, we define n independent random variables, X_1, $X_2, \ldots, X_n$, one for each of the individual trials. To make the example more concrete, suppose we randomly select seven animals from the pen and the outcomes are as listed in the following set:

$$\{F, M, F, F, M, F, M\}.$$

That is, we have the data presented in table 4.1. The expected value of X is stated in equation 4.4.

$$\mu = E(X) = \Sigma X \cdot P(X) = (1)(p) + (0)(1 - p) = p \qquad \boxed{4.4}$$

The formula for the variance of X, noted Var (X) or σ^2, appears in equation 4.5. For our example,

$$\sigma^2 = E(X - \mu)^2 \text{ or } E(X - p)^2 = \Sigma(X - p)^2 \cdot P(X) \qquad \boxed{4.5}$$

it is computed in equation 4.6. Now r, the number of successes in n trials, is expressed by the

$$\sigma^2 = (1 - p)^2 \cdot p + (0 - p)^2 \cdot (1 - p) = p(1 - p) \qquad \boxed{4.6}$$

sum in equation 4.7 and it follows that for $n = 7$, the expected value of r is calculated in equation 4.8.

$$r = X_1 + X_2 + \ldots + X_n \qquad \boxed{4.7}$$

$$E(r) = E(X_1) + E(X_2) + \ldots + E(X_7) = p + p + \ldots + p = np \qquad \boxed{4.8}$$

Similarly, the variance of r is expressed in equation 4.9 and the standard deviation of r is

$$\text{Var}\ (r) = p(1 - p) + p(1 - p) + \ldots + p(1 - p) = np(1 - p) \text{ or } npq \qquad \boxed{4.9}$$

the square root of the variance.

Consider the following example. A farm supply dealer has a bin of parts that we know contains 10 percent defectives. If we consider the selection of a defective part as a success, S, and the selection of a good part as a failure, F, then the expected value of the number of successes, r, per sample of two items is calculated from equation 4.8 as

$$E(r) = np = (2)(0.10) = 0.20,$$

and the variance of r for $n = 2$ is obtained from equation 4.9 as

$$\text{Var}\ (r) = npq = (2)(0.10)(0.90) = 0.18,$$

while the standard deviation is the square root of 0.18 or 0.42.

Poisson Probability Distribution

A second set of problems is described by the small probability of success for any one of several trials of an experiment. Such problems include, for example, applications in crop insurance such as the number of weather-related disasters in a given time period, the number of trucks arriving at a grain elevator during harvest and, hence, the kind of queue or waiting line that forms, the demand for items from a farm supply company's inventory, and the number of defects in a farm machinery manufacturer's product. The Poisson probability distribution is suitable for representing these kinds of problems. It can also be viewed as the limiting form of the binomial as well as a probability distribution in its own right.

As a limiting form of the binomial, it is most useful when the number of trials from a Bernoulli process, n, is large and the probability of a success, p, is small. In this case, the computations we make using the binomial pdf are long and tedious, but they are quite simple with the Poisson pdf. Actually, a Poisson distribution is not described by events with two possible outcomes and a constant probability of success as are Bernoulli trials. However, under certain conditions, we can use the Poisson distribution to solve Bernoulli problems. For example, cultivator frames are continuously manufactured by an extrusion process and the thickness of the wall of the tubing is occasionally below minimum acceptable standards. In a sample of 1,000 feet of such tubing, twenty defective places were found to randomly occur over the tubing, giving an average of two defects per 100 feet. If the probability of finding a specified number of defects per 100 feet of tubing remains constant, we can then consider this Poisson problem as a Bernoulli by viewing a very short length as an independent trial with an outcome of zero or more successes. As long as we keep the mean, np, equal to two defects per 100 feet, we can divide the 100-foot length into 10-foot segments, or 1-foot segments, or whatever is convenient. As the number of pieces, n, becomes large, the probability of more than one defect per piece becomes very small and the difference between the binomial pdf and the Poisson pdf becomes negligible.

Thus, we can say that in a Bernoulli problem, probabilities are associated with the probability, p, of a success in n independent trials and r number of successes. In a Poisson problem, we express probabilities for a given number of successes, l, per unit of space (such as a short length of tubing) and the number of successes, m, in a given amount of space. The letters p and l have the same meaning, and so do r and m. Thus, the expected number of successes, np, for Bernoulli trials corresponds to lm (equation 4.10), the expected number of successes in a given amount of space

$$\mu = lm \tag{4.10}$$

for a Poisson problem.

Consider the following example. In a study of the arrivals of grain trucks at an elevator, the probability of an arrival during any given minute, a trial, is $p =$ 0.0333, while the expected number of arrivals per half hour is

$$np = (30)(0.0333) = 1.$$

By viewing a minute of time as a unit of space, l, and a half hour as the given amount of space, m, then

$$\mu = (0.0333)(30) = 1.$$

We can compute the binomial probabilities for r successes in n trials by our pdf using equation 4.11.

$$P(r \mid n, p) = {}_nC_r\, p^r q^{n-r} \tag{4.11}$$

We know that for small values of p, the resulting probability distribution is highly skewed to the right. For very small values of p and large values of n, as long as np remains constant, the limit of the binomial is the Poisson, which is written in equation 4.12

$$\lim P(r|n,p) = \frac{(np)^r}{r!} e^{-np} \tag{4.12}$$

where e is the base of natural logarithms with the value 2.71828. Now if we use μ in place of np, then we can finally write the formula for the **Poisson** (equation 4.13).

$$P(r|\mu) = \frac{\mu^r}{r!} e^{-\mu} \tag{4.13}$$

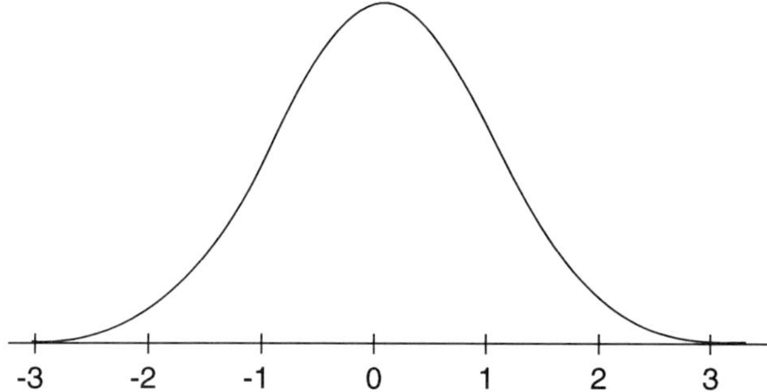

FIGURE 4.2 Standard normal probability distribution

Thus, the Poisson distribution has one parameter, μ, and we can determine the entire distribution once we know it; the $E(r) = \mu$ and $Var\ (r) = \mu$. The standard deviation is the square root of μ. These results come directly from those for the binomial distribution since for it $E(r) = np$ and $Var\ (r) = npq$ and the Poisson is the limit of the binomial distribution as n approaches infinity and p approaches zero, so that $np = \mu$ remains constant. So npq approaches $np = \mu$ as q approaches 1.

Consider the following example. Inspectors in a farm machinery manufacturing plant check 300-foot lengths of hydraulic hose for defects and record the number of defects for each length of hose. From past experience, we know that the number of defects per length of hose follows a Poisson distribution. Suppose we inspect a sample of twenty lengths of hose and observe ten defects. Thus, our estimate of $\mu = 10/20 = 0.5$ is the sample average and the variance. The standard deviation is the square root of 0.5, or 0.7.

The Normal Probability Distribution

While the binomial distribution is the most commonly used probability distribution for discrete variables, the normal is the most helpful continuous distribution. The graph of a general normal distribution looks like the symmetrical, bell-shaped distribution shown in figure 4.2. Its mean, μ, falls at the midpoint of the distribution and its standard deviation, σ, along with μ, completely define the distribution. Since this is the theoretical distribution or true distribution associated with all of the values of the population, we use the Greek letters to refer to the values of the mean and standard deviation rather than $\overline{X}$ and S, which we will use later when we discuss the mean and standard deviation of samples drawn from this distribution.

Some of the characteristics of a normal curve are:

1. The area under the curve between $\mu - \sigma$ and $\mu + \sigma$ is approximately 68 percent of the total area.
2. The area under the curve between $\mu - 2\sigma$ and $\mu + 2\sigma$ is approximately 95 percent of the total area.

3. The area under the curve between $\mu - 3\sigma$ and $\mu + 3\sigma$ is 99.7 percent of the total area.

Thus, while the random variable X theoretically ranges from minus infinity to plus infinity under the normal curve, almost all of the probability is contained in the range of the mean plus or minus three standard deviations.

The location and shape of the normal curve are completely determined by the values of its mean and standard deviation. The value of the mean locates the center of the distribution along the real number line, whereas the value of the standard deviation determines the extent of the curve's spread about the mean. Since all normal curves representing theoretical probability distributions have a total area of one, as the standard deviation increases, the curve must decrease in height and spread out. Because its shape is completely determined by its standard deviation, we may reduce all normal curves to a standard one by a simple change of variable. The easiest normal curve to work with is one with a mean of 0 and a standard deviation of 1 (the standard normal curve, figure 4.2). Thus, when we need to calculate probabilities for any normal curve, we first reduce it to this standard one. The **normal pdf** is written in equation 4.14 and is not easily

$$P(X) = \frac{1}{\sigma\sqrt{2\pi}} e^{\frac{-(X-\mu)^2}{2\sigma^2}} \qquad \boxed{4.14}$$

evaluated; therefore, we use Appendix table 7 to obtain probabilities for the standard normal distribution. This is possible because a point on the x axis of the **standard normal curve** corresponds to a point on the x axis of any normal curve, and we can determine its value by stating how many standard deviations it is away from the mean. For example, suppose we have two curves that differ only by their standard deviations, i.e., the first has mean 0 and standard deviation 1, while the second has mean 0 and standard deviation 3. We can make the second conform to the first by changing the x axis and compressing the curve to one-third its length, i.e., by dividing every term by $\sigma = 3$. Thus, the point $X = 6$ is equivalent to the point $X = 2$ on the standard normal curve because 6 is two standard deviations to the right of the mean, 0. In general, if a point X on the axis of some normal curve with mean μ and standard deviation σ corresponds to a point Z on the standard normal curve, then the point X is Z standard deviations to the right of μ, or $X = \mu + Z\sigma$, which we can rewrite as equation 4.15 by using algebra to solve for Z.

$$Z = (X - \mu)/\sigma \qquad \boxed{4.15}$$

This statement for Z corresponds to the way we standardized individual X values earlier. By this practice of expressing all X values from any given normal curve in terms of corresponding Z values for the standard normal curve, we can reduce all normal curves to the standard one and may use Appendix table 7 for the standard normal for finding probabilities under any normal curve. For example, suppose X belongs to a normal distribution with mean 230 and standard deviation 20, i.e., $X \sim N(230, 20)$. What is the probability of obtaining a value of X from this distribution that is as large or larger than 280?

$$Z = \frac{280 - 230}{20} = \frac{50}{50} = 2.5$$

From Appendix table 7, the probability between the mean and Z is 0.4938. We get the answer by subtracting this value from 0.5, the probability under the curve to the right of the mean, or

$$0.5000 - 0.4938 = 0.0062.$$

Thus, the probability that a value of X from a normal distribution with mean 230 and standard deviation 20 is 280 or larger is 0.0062. Now consider a second question. What is the probability that a value of X from this distribution is smaller than 270?

$$Z = \frac{270 - 230}{20} = \frac{40}{20} = 2.0$$

From the appendix table, the probability between the mean and Z is 0.4772 and the probability to the left of the mean is 0.5. Thus, we find the probability that a value of X from this distribution is smaller than 270 by

$$0.5000 + 0.4772 = 0.9772.$$

Notice that these probabilities correspond to areas under the curve, and for us to sum or subtract probabilities is equivalent to summing or subtracting areas (figure 4.3). However, the probability that X is equal to a single value is not defined, since no area corresponds to a point. We always pose probability questions for a

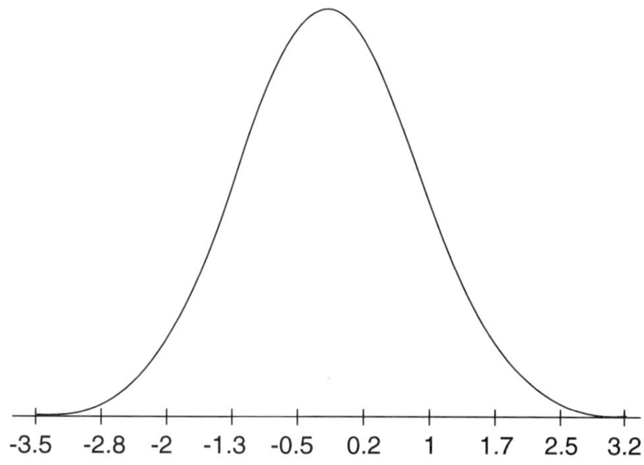

FIGURE 4.3 Standard normal probability distribution

range of values for the random variable X when dealing with continuous distributions such as the normal. In the preceding example, it does not matter, therefore, whether we say "find the probability that X is less than 270" or "find the probability that X is less than or equal to 270." The answer is the same.

Normal Approximation to Binomial Probabilities

When we have Bernoulli trials and want to calculate the probability of r successes in n trials where n is large, sometimes we use the normal approximation to the binomial distribution. For example, if we select a sample of $n = 20$ items randomly from a manufacturing process in which $p = 0.4$, and we want to know the probability of obtaining exactly five defectives, i.e., the probability that $r = 5$, we use the binomial pdf. Thus, we get

$$P(r = 5 \mid n = 20, p = .4) = {}_{20}C_5\ .4^5\ .6^{15} = 0.0746.$$

We can compute the normal approximation of this binomial probability by solving for $E(r)$ and σ_r and using them to find the approximate normal area. We recall that

$$E(r) = np = 20 \cdot 0.4 = 8, \text{ and}$$
$$\sigma_r = (npq)^{1/2} = (20 \cdot 0.4 \cdot 0.6)^{1/2} = 2.19.$$

To find the probability for the r value, we use what at first glance might seem to be trickery, but actually it is not. As we discussed earlier, the normal curve is a continuous curve and we cannot find the probability for a point such as $r = 5$. In this case there is no problem, for if we construct a histogram of the binomial, we see that $r = 5$ is really the midpoint of the class represented by that bar of the histogram and the endpoints of the class, or the bar, are 4.5 and 5.5. Using these endpoints as the values for r, then we find the area under the normal distribution that approximates the area of the bar of the histogram, which is the probability we are seeking. So let's do that. First we standardize each value of r by calculating Z values; next we find the corresponding probabilities from Appendix table 7. Finally we subtract the probabilities to obtain the approximate probability for the area of the bar.

$$Z_1 = \frac{4.5 - 8}{2.19} = \frac{-3.5}{2.19} = -1.60 \qquad Z_2 = \frac{5.5 - 8}{2.19} = \frac{-2.5}{2.19} = -1.14$$

From the appendix table, the probability that

$$Z_1 = -1.60 \text{ is } 0.4452 \text{ and that}$$
$$Z_2 = -1.14 \text{ is } 0.3729.$$

Thus, the

$$P(4.5 \le r \le 5.5) = 0.4452 - 0.3729 = 0.0723.$$

This normal probability closely approximates the true binomial probability of 0.0746. Therefore, we get sufficiently close answers with the normal approximation and we find the calculations are generally easier than with the binomial pdf to justify its use, especially when we want cumulative binomial probabilities. The normal pdf is the limit of the binomial as n becomes very large; consequently the normal approximation improves as n becomes larger. As a rule of thumb, we use the normal approximation anytime the $E(r) > 5$ when $p < 0.5$ and/or $nq > 5$ when $p > 0.5$. The normal approximation is poorer in the tails of the binomial than nearer the mean.

Exercises

1. The Farm Co., a local farm supply business, gives customers who pay in ten days a 1 percent discount, with the full amount due after ten days. In the past, 20 percent of its customers have paid in ten days. In a recent month, The Farm Co. sent out seven invoices. What is the probability that:
   a. Two customers take the discount?
   b. At most two customers take the discount?
   c. No customers take the discount?
   d. All seven take the discount?
   e. Five or more take the discount?
2. Find the mean and variance of the binomial probability distribution in question 1.
3. A produce buyer for Great-Value Supermarkets knows from past experience that 2 percent of the tomatoes received from Mexico are bad and must be discarded. The buyer inspects a shipment that has just arrived by taking a random sample of four tomatoes. What is the probability that:
   a. None are bad?
   b. Two are bad?
   c. At least two are bad?
   d. At most one is bad?
4. A veterinarian knows from past experience that the probability is 30 percent that a calf with a certain disease will recover. The vet is called to a local ranch and diagnoses six calves with the disease. What is the probability that:
   a. All six calves will recover?
   b. At most two will recover?
   c. None will recover?
   d. Draw the probability distribution using a bar chart. Is it symmetrical or skewed? What are its mean and standard deviation?
5. The number of accidents per month in the All Natural Meat Packing Co. is distributed according to the Poisson distribution with a mean of 0.5. During the last month, what is the probability that there were:
   a. No accidents?
   b. One accident?
   c. More than one accident? Hint: Use the complement rule.
6. The number of pickup calls to the Triple A Rendering Co. is distributed according to the Poisson distribution with a mean of three per hour. What is the probability that during the next hour the company receives:
   a. At least two calls?
   b. One call?
   c. No calls?

7. The sales department of The Cotton House, a seed company, makes 500 phone calls a day during its busy season. The probability that any call results in a sale is 0.02. Use the Poisson approximation to the binomial to find the probability that the 500 calls result in:
   a. Exactly ten sales.
   b. More than fifteen sales.

8. For the standard normal distribution, find the probability that Z is:
   a. Less than 1.
   b. Between 1 and 1.5.
   c. Greater than 2.17.
   d. Between 0 and -0.83.
   e. Less than -1.66.
   f. Between -1.2 and 0.34.

9. Given the following probabilities from a standard normal distribution, find the Z value such that:
   a. $P(Z < -z_0) = 0.05$
   b. $P(Z > z_0) = 0.1628$
   c. $P(-z_0 < Z < z_0) = 0.7540$
   d. $P(Z > z_0) = 0.0485$

10. The weight of a fruit dessert packaged by Fruity Co-op is normally distributed with a mean of $\mu = 10$ ounces and standard deviation $\sigma = 0.4$ ounces. If we select a package at random, what is the probability that it weighs:
    a. Less than 10 ounces?
    b. Less than 10.7 ounces?
    c. Between 9.6 and 10.7 ounces?

11. The birth weights of Hampshire pigs are normally distributed with a mean of 4 pounds and a standard deviation of 0.5 pounds. If a pig is randomly selected from a large group of Hampshire newborns, what is the probability that it weighs:
    a. More than 3.5 pounds?
    b. More than 5 pounds?
    c. Between 4.5 and 5 pounds?
    d. Between 3.5 and 4.5 pounds?

12. If the number of blooms on 6-inch pots of chrysanthemums is normally distributed with a mean of 16 and a standard deviation of 2, what is the probability that a pot selected at random from a large greenhouse full of these pots will contain:
    a. Less than fourteen blooms?
    b. Less than twenty blooms?
    c. Between twelve and twenty blooms?
    d. 90 percent of the pots will contain more than _____ blooms.

CHAPTER

Sampling and Sampling Distributions

In statistics, we generally make inferences about variables of interest in the population based on analysis of sample data. Thus, samples are important in statistics. Of course, the sample must be representative of the population if the inferences based on the statistics calculated from the sample are to be correct. There are two types of sampling procedures that we can use to select samples—probability-based procedures and convenience sampling. We examine these in turn.

Probability-Based Samples

A **probability-based** sample is selected if we set up the sampling procedure so that every element in the population has a known chance of being included in the sample. If we use this type of sampling procedure, then we can include probability statements with the analysis of the sample data. When we decide to take the sample itself, there are several methods available. Each leads to a different kind of sample. Our choice of method depends upon several factors:

1. *Whether the population is homogeneous.* Since sampling involves error, the type of sample design we use depends upon the characteristics of the population variable being sampled. If the variable or attribute is approximately evenly divided among members of the population, then we have a homogeneous population and we can use a simple sample design such as random or systematic sampling. If, on the other hand, the attribute is concentrated at different levels among different groups of the population, then we try to identify the groups and sample each group separately. This leads to more complicated sample designs such as

stratified or cluster sampling that allow us to account for the differences in the population that might cause the variable we are trying to measure to have different values.

2. *The degree of accuracy required.* Other things being equal, we want to select a sample design that will make the variance of the sample statistic as small as possible for the size of sample we have chosen. Otherwise, we have to increase the sample size to reduce the sampling error. **Sampling error** is defined as the difference between the value of the statistic or sample variable and the value of the population parameter or population variable. This difference arises because of chance or random variations in the selection of elementary units for the sample. Thus, the greater the variance of the sample statistic, the greater the sampling error, or the less reliable the sample.
3. *The cost of the sampling plan.* Although we can make the sample more reliable by increasing its size, the cost associated with the sampling procedure also increases with its size. Cost of sampling is usually directly related to sample size, whereas reliability is usually indirectly related. Usually, sample size must be increased considerably to get a small increase in reliability. So we have the concept of **efficiency.** A sample design is more efficient than another if it results in lower costs but the same degree of reliability. Thus, cluster sampling is usually more efficient than simple random sampling when the population to be sampled is scattered over a wide area and considerable travel costs would be involved in drawing the simple random sample.

By sampling design, we mean the design of probability sampling. Thus, let's look at a few of the more commonly used designs and evaluate each in terms of its reliability and efficiency.

Simple Random Sampling

In simple random sampling, every element in the population available for sampling has an equal probability of being selected. Also, if we randomly choose a sample of size n, every possible sample of that size has the same probability of being drawn. Randomness is the main factor to consider in the design of a simple random sample. But achieving randomness is not always simple. If, for example, we want to survey the opinions of students at a large university about agriculture, we cannot select a **random sample** simply by interviewing students at the door of the student union until we get enough interviews. Many students may never go near the place and, thus, there is zero probability of interviewing them. A sample obtained in this manner is not a random sample.

We frequently use the "thorough mixing" method to assure randomness in a sample, in which we number all of the items in the population or otherwise identify them so that every item to be sampled is no different from any other one. We then place the numbers in a large container and mix them thoroughly.

Next, we specify the size of the sample and draw one unit at a time from the container until we obtain the desired number. If we are sampling with replacement, we mix the items thoroughly each time before we draw so that the probability of selecting an item remains the same. For sampling without replacement, the probability increases for the remaining items being drawn because we keep reducing the total available by our selection process. But all of the remaining items have an equal chance of being drawn, which keeps the probability of obtaining a sample of size n the same as for any other combination of items we could draw from that population.

We can simplify our sampling procedure by using a table of random numbers like that in Appendix table 1. The table contains positive integers numbered from 0 to 9 and can be constructed using the method of thorough mixing. We place ten cards with the numbers from 0 to 9 in a container, draw one out, record its value as the first digit in the random number table, replace the card, draw again, and enter the second card's value as the second digit in the table, and so on until we fill the table. Perhaps a more efficient procedure is to use a computer with a random number generating program.

In using the table, we can combine any two numbers to get random numbers from 00 to 99, and three to obtain numbers from 000 to 999, and so on. In general, we select an appropriate number of digits for the size of the population we are sampling. For example, if we have a population of 10,000 items and want to select a sample of size 100, we must first make a list of the population and assign a number from 0000 to 9999 to each element. Then, we choose a random starting place in the table and read it across, down, or diagonally, and select successive numbers until we obtain 100. We throw out duplicated values and replace them with new ones. Finally, we go to the population and select the numbered items for the sample corresponding to those we selected from the table.

We can use a simple random sample if we can easily identify the sampling units and if the population is small and homogeneous. With a large population, the method is expensive and time-consuming because we must number all of the items. Also, if the items close to one another are more homogeneous than those farther away, as in a survey of family incomes, simple random sampling may not provide a representative sample.

Systematic Sampling

We frequently use a **systematic sample** rather than a simple random if we have access to a list of the population. With this procedure, we obtain the sample by taking every kth unit in the population, where k stands for some whole number that is approximately the sampling ratio N/n. Thus, if $N = 10{,}000$ and $n = 500$, then $k = N/n = 10{,}000/500 = 20$, and we select the sample by taking every twentieth unit in the population. For every unit in the population to have an equal chance of being selected, we must choose a random starting point. Thus, in the preceding example, we randomly select a number between 1 and 20. If it is 8, then the sample consists of items 8, 28, 48, etc. until we select all 500 values. Systematic sampling works well when we have a list of the population readily available, such as a membership directory, a list of farmer borrowers from a certain bank, etc. We occasionally use systematic samples in statistical quality control, especially when we are sampling from a current production stream at regular intervals.

With the systematic random sampling procedure, we select a more representative sample than with simple random sampling if the units in the population resemble other nearby items more than they resemble items far away, as in income data. On the other hand, if a hidden cycle or period is in the data with the same amplitude as the sampling interval, systematic sampling may be a poor choice. For example, we do not use systematic sampling for investigating daily grocery store sales, especially if we sample every seventh day's sales, because grocery sales are not evenly distributed throughout the week. Thus, if we choose Tuesday, we miss all of those shoppers who go on the weekend to take advantage of specials or time off from work, and our estimate is not correct.

Stratified Sampling

If we know something in advance about the population and use it to partition items into various layers or strata, then **stratified sampling** is more efficient than simple random sampling. We choose the strata so that they contain similar characteristics and are more homogeneous than the population as a whole. We then sample each stratum using simple random sampling and form the overall sample with subsamples from each stratum. There are several ways to decide how to select the number of items to include in each subsample. One way is to use a proportionate stratified sample, which assures that the number in the sample from each stratum is in the same proportion as the number of items in the population for that stratum. For example, we might sample cotton yields using soil classes as strata. Another way is to make the sampling ratio proportional to the standard deviation of a stratum so that the more homogeneous the stratum, the smaller its standard deviation and, thus, the smaller the proportion we include in the sample from that stratum. This makes a disproportionate stratified sample, for example, when we sample costs and use size of farm as strata. In this case, we select a much smaller proportion of the small farms than of the large farms.

By stratification, we set up classes so that the sampling units within each class are approximately uniform, but the classes are different. We control the proportion of the overall sample from each stratum and do not leave it to chance, therefore assuring a representative sample. If the variance of the observed characteristic within each stratum is smaller than that for the entire population, as is usually true, the reliability of the sample for a given size will be better or the efficiency for a given degree of reliability will be greater because we used a stratified sample. Thus, we find stratification useful in handling heterogeneous populations. In some cases, we must construct several layers, or substrata, to obtain full benefit from stratification. For example, we may stratify cotton yields according to whether the land is dry or irrigated, and then stratify soil classes within those categories.

Cluster Sampling

Cluster sampling is diametrically opposed to stratified sampling. With cluster sampling, we first select groups of individual items called clusters from the population at random, and then choose all or a subsample of the items within each cluster to make up the overall sample. For best results, we want differences between clusters to be as small as possible, and differences among individual items within each cluster to be as large as possible. We want the cluster to be a miniature of the population, so that any cluster is a representative sample. In practice, such requirements may be difficult to meet, but we must keep them in mind when designing a cluster sample.

Clusters are often called primary sampling units. If all items in the selected clusters are included in the sample, we have single-stage sampling. If we take a random subsample of the items in the cluster, we have two-stage sampling. There is also multi-stage sampling. For example, in an attitude survey of all U.S. college students about agriculture, we may select universities as clusters or primary sampling units, schools within universities as the second stage, and students within the schools as the third stage.

The chief advantage of cluster sampling is its low cost for a given degree of reliability. Thus, when we sample from a geographically distributed population, as in the attitude survey example, we take geographical areas as the primary sampling units. The variance of a cluster sample tends to be greater than that for the other sample designs we discussed, but its lower costs allow us to take a much larger sample for less money. And the larger sample reduces the variance enough to cause cluster sampling to be more efficient per dollar spent.

Sequential Sampling

Sequential sampling is widely used in quality control and can be useful in agricultural budgeting studies. It involves testing a relatively small sample in quality control and, on the basis of the sample outcome, deciding whether to accept or reject the whole lot. If the small sample does not lead to a clear decision, we increase the sample size, i.e., we sample more units until we can reach a decision. This procedure keeps sampling costs low since many times we can make decisions on the basis of results from a very small sample. In budgeting studies, we interview a small group of farmers concerning output practices. If they agree, then we construct the budget on the basis of those practices. If not, we take a larger sample until there is agreement.

Nonprobability Samples

When we use nonprobability samples, all of the items in the population do not have a known chance of being included in the sample. We employ such procedures for cost or efficiency reasons, but they do not provide the objectivity offered by probability sampling designs.

Convenience Sampling

In **convenience sampling,** we select a sample based on our convenience and without regard to whether the population is adequately represented. It is convenient in terms of time, cost, and/or administration. We often see this type of sampling on local TV news broadcasts when a reporter interviews the "person on the street" to obtain his or her opinion about some topic in the news. This design should not be used to make inferences about populations.

Judgment Sampling

With **judgment sampling,** we use our judgment rather than some objective method such as random numbers for selecting elements of the population for the sample. For example, the local county extension agent may be called upon by a researcher to hand-pick farmers to interview about management practices because the researcher wants to talk only to the "best" managers.

Quota Sampling

In **quota sampling,** we construct the sample to look like the population with regard to some major attribute. We draw elements of the population until we reach the quota for that aspect of the attribute, and then we choose elements containing a different aspect, and so on. For example, if we know that a county contains 20 percent wheat farmers, 60 percent cotton farmers, and 20 percent livestock farmers, we can select a quota sample that contains exactly those same percentages. Because these sample estimates are subject to greater variability than that for probability sampling, we recommend the probability sampling procedures.

Sampling Distributions

A sampling distribution may help us understand problems related to sampling. With a sampling distribution, we can consider how such statistics as the mean or the variance may vary from sample to sample if we draw repeated simple random samples of the same size from the same population. The probability distribution of such a statistic is called its **sampling distribution.** Thus, we have a sampling distribution of the mean, a sampling distribution of the variance, and so on. These sampling distributions are a basic concept of statistical inference and are also important in statistical decision theory.

Sampling Distribution of the Mean

The idea of a sampling distribution is based on repeated sampling. If we take repeated samples of the same size from a population, and calculate the arithmetic mean of each sample, we can treat those values of the arithmetic mean as a random variable. That is, the arithmetic mean varies in value depending upon which elements of the population we randomly select for the sample. If we treat all of the

TABLE 5.1 Sample Means for All Possible Samples of Size Two

Sample Combination	Sample Measurement	Total (ΣX)	Sample Mean ($\bar{X}$)
X_1, X_2	90, 80	170	85
X_1, X_3	90, 100	190	95
X_1, X_4	90, 80	170	85
X_1, X_5	90, 90	180	90
X_1, X_6	90, 100	190	95
X_2, X_3	80, 100	180	90
X_2, X_4	80, 80	160	80
X_2, X_5	80, 90	170	85
X_2, X_6	80, 100	180	90
X_3, X_4	100, 80	180	90
X_3, X_5	100, 90	190	95
X_3, X_6	100, 100	200	100
X_4, X_5	80, 90	170	85
X_4, X_6	80, 100	180	90
X_5, X_6	90, 100	190	95

random samples as equally likely, then we can place the various values of the arithmetic mean into a probability distribution by adding the probabilities associated with each value. The resulting probability distribution has a certain shape, regardless of the shape of the distribution of the population from which we draw the samples. Also, the expected value of this probability distribution equals the value of the population mean, and its variance has a particular relationship with the population variance. This can best be illustrated by an example.

Suppose a farmer has a new young bull that in his first year sires only six calves with birth weights of

$$X = \{90, 80, 100, 80, 90, 100\}$$

pounds. The mean, μ, of this population is 90 pounds and the variance, σ^2, is 66.67. If we randomly draw all of the different possible samples of size two from this population, there are

$$_6C_2 = 6!/(6 - 2)!(2!) = 15$$

of these. For each sample, we calculate the sample mean and use the sample mean as a random variable to construct a sampling distribution of the mean (table 5.1). From the table, we see that the possible values are closely grouped around the population mean. Since the sample means have frequencies of occurrence, we can place them in a probability distribution, as shown in table 5.2.

TABLE 5.2 Probability Distribution of Sample Means

Sample Mean ($\overline{X}$)	Frequency (f)	Probability, $P(\overline{X})$	$\overline{X} \cdot P(\overline{X})$
80	1	1/15	80/15
85	4	4/15	340/15
90	5	5/15	450/15
95	4	4/15	380/15
100	1	1/15	100/15
Totals	15	15/15 = 1	1,350/15 = 90

We can calculate the mean of the probability distribution in table 5.2 by using the expected value formula, i.e.,

$$E(\overline{X}) = \Sigma\overline{X} \cdot P(\overline{X}) = (80)(1/15) + (85)(4/15) + (90)(5/15) + (95)(4/15) + (100)(1/15) = 1{,}350/15 = 90,$$

which is the same as the value of the population mean, μ. This relationship always holds and is formally stated by the Central Limit Theorem.

Central Limit Theorem

If random samples of size n observations are drawn from a population with finite mean μ and standard deviation σ, then the sample mean $\overline{X}$ is approximately normally distributed with mean μ and standard deviation $\sigma/\sqrt{n}$ when n is large. As n increases in size, the approximation becomes more accurate.

If we plot the probability distribution for the small population of calf birth weights, it is rectangular in shape (figure 5.1), but the sampling distribution of the mean, which we obtain by plotting all possible sample mean values against their probabilities, is more shaped like the normal distribution (figure 5.2).

Standard Error of the Mean

The standard deviation of the sampling distribution of the means is called the **standard error of the mean** so that it is not easily mixed up with the population standard deviation. The standard error is related to the population standard deviation as stated in the Central Limit Theorem, i.e., it is equal to the population standard deviation divided by the square root of the sample size. When we have small populations, as in the calf birth weight example, we have to apply the **finite population correction factor** to the formula. Thus, we have equation 5.1, which contains this factor.

$$\sigma_{\overline{X}} = \frac{\sigma}{\sqrt{n}}\sqrt{\frac{N-n}{N-1}} \qquad \boxed{5.1}$$

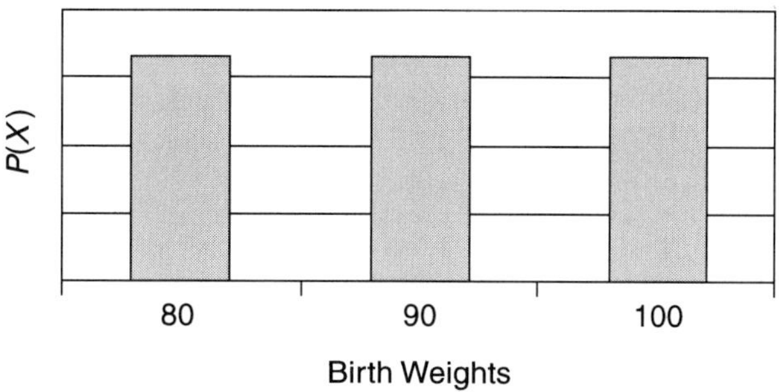

FIGURE 5.1 Probability distribution for the population of calf weaning weights

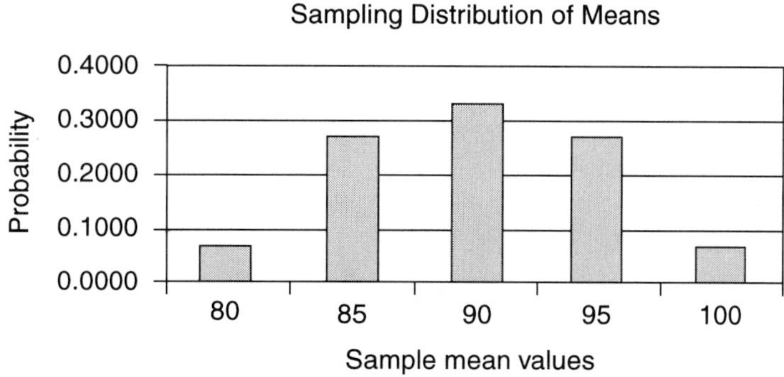

FIGURE 5.2 Sampling distribution of the sample means for samples of size two

where σ = population standard deviation,
N = population size,
n = sample size, and
term under right square root = finite population correction factor.

Using formula 5.1 with the calf birth weight example, we get

$$\sigma_{\overline{X}} = \frac{8.167}{\sqrt{2}}\sqrt{\frac{6-2}{6-1}} = 5.78 \cdot 0.8944 = 5.17$$

We see that the correction factor evaluates to approximately 0.89 and thus reduces $\sigma_{\overline{X}}$ somewhat. Alternatively, we could have calculated $\sigma_{\overline{X}}$ directly from

TABLE 5.3 Calculation of the Standard Error of the Mean Using the Probability Distribution

Sample Mean, $\bar{X}$	$P(\bar{X})$	$\bar{X} - \mu$	$(\bar{X} - \mu)^2$	$(\bar{X} - \mu)^2 \cdot P(\bar{X})$
80	1/15	$80 - 90 = -10$	100	100/15
85	4/15	$85 - 90 = -5$	25	100/15
90	5/15	$90 - 90 = 0$	0	0
95	4/15	$95 - 90 = 5$	25	100/15
100	1/15	$100 - 90 = 10$	100	100/15
Totals	$15/15 = 1$			$400/15 = 26.67 = \sigma_{\bar{X}}^2$
				$\sigma_{\bar{X}} = \sqrt{26.67} = 5.17$

the data in table 5.2. The formula for the variance of a probability distribution is given by equation 5.2, which we can apply to the table

$$\sigma^2 = \Sigma(X - \mu)^2 \cdot P(X). \quad \boxed{5.2}$$

data. These calculations are presented in table 5.3 and result in the same answer, 5.17. But, if we know the value of the population standard deviation, we use the finite correction factor formula because it is quicker. In fact, no one generally takes all possible samples from any distribution—we usually draw only one sample. Thus, we do not have the option of computing $\sigma_{\bar{X}}$ by means of the probability distribution as in table 5.3. Fortunately, we have the formulas based on σ.

Exercises

1. What is the difference between a census and a sample? Explain. Provide three or more reasons why sampling may be more desirable than a census in certain cases.
2. Give an example in which the population is:
   a. Infinite.
   b. Considered infinite, but is actually finite.
   c. Finite.
3. What is the purpose of having a control group in a study? Provide an example of a situation in which a control group would be useful.
4. What is the difference between random and systematic errors? Which can and should be eliminated in sampling? Explain.
5. Using the table of random numbers, Appendix table 1, select a simple random sample of ten items from a population of 100. Describe the selection process used.
6. Using a table of random numbers, Appendix table 1, select a 5 percent systematic sample from a population of 300. Describe the selection process used.
7. Repeat exercises 5 and 6 using the Sampling procedure from the Tools: Data Analysis Menu in Excel. Print your results and explain the selection process.
8. A national farm co-op has salespeople in Sacramento, Minneapolis, Houston, and Atlanta. Each office has 16 salespeople. The weekly sales of any salesperson are normally distributed with a mean of $30,000 and a standard deviation of $6,400. What range about the mean contains 95.5 percent of:
   a. A randomly selected salesperson's weekly sales?
   b. The average weekly sales per salesperson in the Houston office?
   c. The average weekly sales per salesperson in the co-op?
9. What is the probability of drawing a simple random sample with a mean of 50 or more from a population with a mean of 46 if the sample size is 81 and the population variance is 324?
10. The average weekly food purchase of 50 people is $120, with a standard deviation of $25. If a randomly selected group of five people from this population is to be surveyed about buying practices, what is the average food purchase for the possible groups of five? The standard deviation?

11. A plant geneticist has five plots of new varieties of cotton that yield as follows:

Plot	Yield per Acre
V	520
W	500
X	540
Y	570
Z	480

Two plots are to be selected at random to estimate the average yield.

a. List all possible samples of two plots, and calculate the mean yield for each sample.

b. Calculate the mean and variance of the sampling distribution from the data in part a. and also from the original data.

12. Cal Co-op, a large group of orange growers on the West Coast, has decided to construct a juice processing plant. It has randomly selected sixteen similar plants and estimated their average gallons of juice per day as 3,000 with a standard deviation of 400 gallons. What is the probability that a randomly selected plant will have a daily average within 200 gallons of the group mean?

CHAPTER

Introduction to Statistical Inference: One Sample

In statistics, we usually deal with samples, even though we are primarily interested in the populations from which the samples are selected. Our purpose is to make statements about the value of some variable in the population based on the sample results. As we stated earlier, a random sample is actually a random subset of the population and our concern is in making inferences about the population. Our problem is that we do not know whether the inferences we make are correct because of the variability in the data on which we base our conclusions and because of the variability among samples. To keep from being wrong most of the time, we use probability theory as a basis for our inferences. Thus, we define **statistical inference** as a judgment about some variable in the population based on probability theory and sample data.

Statistical inference involves making estimates of population variables and testing hypotheses about their behavior. We call population variables **parameters** and sample variables **statistics** so that we readily know which set we are considering. Thus, we always calculate statistics from sample data and use them as the basis for estimating parameters. More generally, we name the random variables we select to estimate population parameters **estimators,** while we call specific values of these random variables **estimates** of the population parameters. We present these estimates in two different forms—as points and as intervals. A **point estimate** is simply one value. For example, we can estimate the population mean from a set of sample data as equal to 2. On the other hand, we can construct an interval such as 1–3 and state that the population mean falls somewhere within it. Which type of estimate is best depends upon what we must have to answer the problem at hand and on the probability statement attached to it.

Properties of Estimators

The population mean and variance are the parameters we most often estimate since they define most probability distributions. If we choose to estimate the population mean, what statistic would we select? There are several estimators to choose from, but the best is $\bar{X}$, the sample mean. How do we know that it is best? It turns out that there are several desirable properties of a good estimator or sample statistic and the sample mean satisfies them all. We now look at these more closely.

Unbiasedness

We have an unbiased estimator if its expected value equals the population parameter; i.e., if $E(\bar{X}) = \mu$ for the sample mean. Or in general, if $\hat{\eta}$ (the Greek letter eta with a ˆ (hat) on it, read eta-hat) is the estimator or sample statistic and η is the population parameter, $\hat{\eta}$ is unbiased if $E(\hat{\eta}) = \eta$, i.e., if its sampling distribution is centered directly over η. Thus, $\hat{\eta}$ is biased if its expected value is different from η. Thus, we define **bias,** B, in equation 6.1. A good estimator

$$B \equiv E(\hat{\eta}) - \eta. \quad \boxed{6.1}$$

should be unbiased, and $\bar{X}$ is unbiased since we learned earlier that the mean of the sampling distribution of $\bar{X}$ is equal to μ, the population mean, when we discussed the sampling distribution of the mean.

Efficiency

As well as being on target, on the average, we also want the sampling distribution of the estimator to be concentrated, that is, to have a small variance. This is the idea of *efficiency.* Thus, if there are two estimators, $\hat{\eta}$ and $\acute{\eta}$ (eta-prime) of η, then $\hat{\eta}$ is more efficient than $\acute{\eta}$ if the variance of the sampling distribution of $\hat{\eta}$ is smaller, i.e., if its sampling distribution is more peaked. An estimator like $\hat{\eta}$, which is more efficient, will provide a point estimate that is closer to our target, the true value of the parameter η. Or in the case of an interval estimate, it provides a smaller interval and, hence, a more precise one for the same size sample. Of course, by increasing the sample size, n, we can reduce the variance of some estimators. Thus, another way of looking at the greater efficiency of $\hat{\eta}$ is to recognize that $\acute{\eta}$ will produce an accurate point or interval estimate only if we take a larger sample. Hence, using $\hat{\eta}$ is more efficient because it costs less to sample. A useful relative measure of the efficiency of two unbiased estimators is the ratio of their variances (equation 6.2).

$$\text{Relative efficiency of } \hat{\eta} \text{ compared to } \acute{\eta} \equiv \text{var}(\acute{\eta})/\text{var}(\hat{\eta}) \quad \boxed{6.2}$$

Sufficiency

The properties of unbiasedness and efficiency are desirable for an estimator, particularly with small samples. Another property is **sufficiency.** A sufficient estimator, $\hat{\eta}$, uses all of the information about the population parameter, η, that is contained in the sample data. Thus, a sufficient estimator uses all of the sample observations in its calculation. The median, Md, is not a sufficient estimator because it uses only a ranking of the observations and not their values, but $\bar{X}$ and S^2 are sufficient estimators. Sufficiency is a necessary condition for efficiency.

Consistency

A **consistent estimator,** $\hat{\eta}$, is one that concentrates completely on its target as sample size increases indefinitely toward infinity. As sample size becomes infinite, a consistent estimator $\hat{\eta}$ provides a perfect point estimate of the target η. Just as the variance is a good measure of the spread of a distribution about its mean, so the mean squared error (MSE) is a good measure of how the sampling distribution of an estimator $\hat{\eta}$ is spread about its target value η. Consistency requires that MSE be zero in the limit, as stated in equation 6.3.

$$\text{MSE} \equiv E(\hat{\eta} - \eta)^2 \text{ and } \hat{\eta} \text{ is consistent as } n \rightarrow \infty \text{ if MSE} \rightarrow 0. \tag{6.3}$$

Equation 6.4 expresses the relationship between mean squared error, bias, and the variance. Therefore, $\hat{\eta}$ is a consistent

$$\text{MSE} \equiv E(\hat{\eta} - \eta)^2 = [\text{Bias } \hat{\eta}]^2 + \text{var } (\hat{\eta}) \tag{6.4}$$

estimator if and only if its bias and variance both approach zero as n approaches infinity. If only the bias approaches zero, we call the estimator asymptotically unbiased—a condition that is weaker than consistency.

Consistency by itself does not guarantee that an estimator is a good one. For example, as an estimator of μ in a normal population, the sample median Md is consistent since it is unbiased and its variance approaches zero as sample size increases toward infinity. But it is not as good as the sample mean, $\bar{X}$, because $\bar{X}$ is both consistent and efficient, i.e, its variance is always less than the median's. Thus, we want an estimator with all of these properties; and we have them with $\bar{X}$ and S^2. Thus, these two estimators are the commonly used point estimators of μ and σ^2.

Interval Estimation

An estimate of a population parameter does not have to be a single value; instead it can cover a range of values. We call estimates that specify a range of values interval estimates or **confidence intervals.** We construct confidence intervals for population parameters such as μ and base them on the correct sampling distribution of the estimator, such as the sampling distribution of $\bar{X}$. If, for example, we use $\bar{X}$ to make a confidence interval for μ, then we want to attach a probability measure to the interval. That is, we want to say something like "we are certain that if we construct intervals like this one around $\bar{X}$ values obtained from repeated samples of size n from the population, then 100 $(1 - \alpha)$ percent of them contain μ." We use the probability measure $(1 - \alpha)$ where α (the Greek letter alpha) is a very small number, usually less than 0.10, and we select it based on the cost of being wrong. The greater the cost, the smaller the number. Thus, if $\alpha = 0.05$, we construct a 95 percent confidence interval.

Notice that in our statement interpreting the confidence interval, we refer to repeated samples of size n. In actual practice, we select only one sample, compute the value of the estimator (such as $\bar{X}$), select the degree of confidence (choose $1 - \alpha$), construct the interval, and behave as if the parameter (μ) is contained within it. And why not? Unless the value of the estimator lies in the extreme end of one of the tails of the sampling distribution, the interval actually contains the parameter. And we know the probability of getting a value that lies in the tails of the sampling distribution; it is α, which we set to be small from the beginning.

The width of a confidence interval depends upon the values we select for the sample size n and for α. The smaller the size of sample, the wider the interval, because the variability of the sampling distribution is related to n ($\sigma_{\bar{X}} = \sigma/\sqrt{n}$). The larger the variability, the more spread out the sampling distribution and, hence, the wider the confidence interval. Also, the smaller the size of α selected, the wider the confidence interval. The larger the interval, the greater the range within which the parameter falls. Thus, we must weigh the cost of additional sample observations and the cost of greater confidence in constructing the interval.

The last thing that influences the size of a particular confidence interval is the sampling distribution of the statistic we use for estimating the parameter. Generally, the procedure for obtaining a confidence interval for a parameter η is first to find a point estimate $\hat{\eta}$ of this parameter. Then we construct the confidence interval by finding the interval of values around $\hat{\eta}$, which according to its sampling distribution, yields the desired degree of confidence. Since we use different sampling distributions to estimate the population mean, the binomial parameter p, and the population variance, we describe the procedure for constructing confidence intervals for each of these in turn.

Confidence Intervals for μ with σ Known

To construct a confidence interval for μ, suppose we draw a random sample from a normally distributed population with known standard deviation, σ. The sample statistic for estimating μ is $\bar{X}$ because it has all the desirable properties of an estimator discussed earlier. The sampling distribution of $\bar{X}$ is the normal distribution with mean $E(\bar{X}) = \mu$ and standard error $\sigma_{\bar{X}} = \sigma/\sqrt{n}$. The variable Z as expressed in equation 6.5 has a standardized normal distribution,

$$Z = (\bar{X} - \mu)/\sigma_{\bar{X}} \tag{6.5}$$

i.e., $Z \sim N(0,1)$, and we can find probabilities for Z from Appendix table 7. Now suppose we let Z_α represent a value of Z in the right tail of the distribution such that the probability of observing values of Z greater than Z_α is α, i.e., $P(Z > Z_\alpha) = \alpha$. Also, let $-Z_\alpha$ be a point in the left tail of the distribution such that $P(Z < -Z_\alpha) = \alpha$. Two such points are illustrated in figure 6.1. Rather than to find points cutting off α in the tail of the standardized normal distribution, it is more appropriate to locate points that cut off $\alpha/2$ probability in each tail (so that the total area equals α). To do this, let

$Z_{\alpha/2}$ represent the value such that the probability $P(Z > Z_{\alpha/2}) = \alpha/2$ and let $-Z_{\alpha/2}$ equal the point such that $P(Z < -Z_{\alpha/2}) = \alpha/2$. For example, if $\alpha = 0.05$, then $\alpha/2 = 0.025$ and the value of $Z_{\alpha/2}$ that cuts off 0.025 probability in the tail of the standard normal distribution is, from Appendix table 7, equal to 1.96. Since the normal distribution is symmetric, the value of $-Z_{\alpha/2}$ must be -1.96. The probability that Z falls between the two limits ± 1.96 is

$$P(-1.96 < Z < 1.96) = (1.0 - 0.05) = 0.95,$$

which is what we want for a 95 percent confidence interval. So, in general, the probability $(1 - \alpha)$ is given by equation 6.6. This is a $100(1 - \alpha)$ percent confidence interval for Z. However,

$$P(-Z_{\alpha/2} < Z < Z_{\alpha/2}) = (1 - \alpha). \qquad \boxed{6.6}$$

we want to construct a confidence interval for μ, not Z. Now we can algebraically manipulate equation 6.6 to get what we want. We begin with equation 6.5, which defines Z, and substitute this expression for Z in the probability statement in equation 6.6, and then rearrange terms to provide the appropriate confidence interval for μ. After the substitution we have

$$P(-Z_{\alpha/2} < (\bar{X} - \mu)/\sigma_{\bar{X}} < Z_{\alpha/2}) = (1 - \alpha).$$

Now multiply both sides by $\sigma_{\bar{X}}$ to get

$$P(-Z_{\alpha/2} \cdot \sigma_{\bar{X}} < (\bar{X} - \mu) < Z_{\alpha/2} \cdot \sigma_{\bar{X}}) = (1 - \alpha).$$

Next add $(-\bar{X})$ to both sides, which gives

$$P(-\bar{X} - Z_{\alpha/2} \cdot \sigma_{\bar{X}} < -\mu < -\bar{X} + Z_{\alpha/2} \cdot \sigma_{\bar{X}}) = (1 - \alpha).$$

Finally, multiply both sides by (-1), which changes the direction of the inequalities and gives us the confidence interval on μ we are seeking, equation 6.7.

$$P(\bar{X} + Z_{\alpha/2} \cdot \sigma_{\bar{X}} > \mu > \bar{X} - Z_{\alpha/2} \cdot \sigma_{\bar{X}}) = (1 - \alpha) \qquad \boxed{6.7}$$

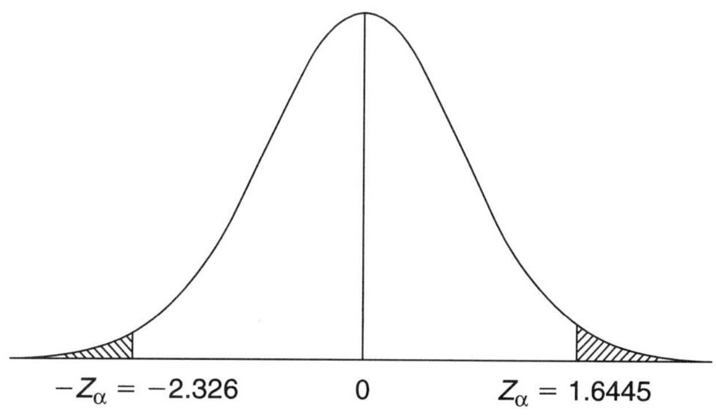

FIGURE 6.1 Sampling distribution of the mean

To illustrate the use of equation 6.7, assume that the population of corn yields for a farming region is normally distributed with $\sigma = 6.2$ bushels and that we take a random sample of size $n = 36$ farms and compute the average yield as $\bar{X} = 132$ bushels. Construct a 95 percent confidence interval on μ, the true average corn yield for the region. Since we want a 95 percent confidence interval, the two appropriate Z values are $Z_{\alpha/2} = 1.96$ and $-Z_{\alpha/2} = -1.96$. We recall that

$$\sigma_{\bar{X}} = \sigma/\sqrt{n} = 6.2/\sqrt{36} = 6.2/6 = 1.033.$$

We substitute these values into equation 6.7 for the confidence interval to obtain:

$$P(\bar{X} + Z_{\alpha/2} \cdot \sigma_{\bar{X}} > \mu > \bar{X} - Z_{\alpha/2} \cdot \sigma_{\bar{X}}) = (1 - \alpha)$$
$$P(132 + (1.96)(1.033) > \mu > 132 - (1.96)(1.033)) = 0.95$$

or

$$P(132 + 2.03 > \mu > 132 - 2.03) = 0.95$$

or, with rounding,

$$134 > \mu > 130.$$

We act as though the average corn yield is between 130 and 134 bushels because if we take repeated samples of size 36 from this population and calculate confidence intervals for each sample mean, then ninety-five out of 100 of such samples provide sample means close enough to μ so that the intervals actually contain μ. We have no reason to believe that the sample we selected is any different from one of those ninety-five. Of course, we could be unlucky and draw an atypical sample that would not be one of the ninety-five, but the odds are so low we discount that. Nevertheless, we state the probability associated with the confidence interval so that our audience understands the odds.

For the confidence interval we calculated in the previous example, the probability of making an error by assuming that μ is in the interval is $\alpha = 0.05$. If we want a smaller risk of error, then we must use a wider interval. For example, if we have an error of $\alpha = 0.01$, or a 99 percent confidence interval, we get the appropriate Z values from Appendix table 7: $Z_{\alpha/2} = 2.57$ and $-Z_{\alpha/2} = -2.57$ (for $\alpha/2 = 0.005$). Using equation 6.7, the new 99 percent confidence interval is:

$$P(\bar{X} + Z_{\alpha/2} \cdot \sigma_{\bar{X}} > \mu > \bar{X} - Z_{\alpha/2} \cdot \sigma_{\bar{X}}) = (1 - \alpha)$$
$$132 + (2.57)(1.033) > \mu > 132 - (2.57)(1.033)$$
$$132 + 2.65 > \mu > 132 - 2.65$$
$$134.6 > \mu > 129.4$$

Since this interval is wider than the previous one, we have greater confidence that it contains the parameter μ.

For *non-normal populations* in which we know σ and have small n (less than 30), the distribution of $Z = (\bar{X} - \mu)/\sigma_{\bar{X}}$ is not normal and we have no convenient procedure for constructing a confidence interval. However, if n is large ($n \geq 30$), the Central Limit Theorem tells us that the sampling distribution of Z is ap-

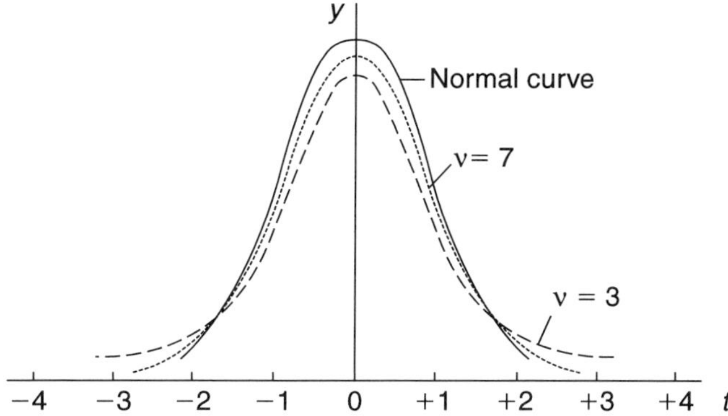

FIGURE 6.2 A *t* distribution for selected values of nu

proximately normal; therefore, we can use the formula for a confidence interval that we have been using.

Confidence Intervals for μ with σ Unknown

In the previous section, we assumed that we knew the population standard deviation and we wanted to estimate the population mean using sample data. It is more common, however, that if we must estimate the mean, we must also estimate the standard deviation from sample data. Under these circumstances, we use S to approximate σ, and equation 6.8 has a t distribution with

$$t = (\bar{X} - \mu)/(S/\sqrt{n}) \tag{6.8}$$

$\nu = (n - 1)$ **degrees of freedom** (assuming the population distribution is normal). The t distribution is a family of pdfs whose shape depends upon ν (the Greek letter nu used to designate degrees of freedom) and not upon σ, the population standard deviation. This distribution is bell-shaped like the normal distribution but is not quite as peaked, thus it is fatter in the tails (figure 6.2). Since ν determines its shape and ν increases as the sample size increases, the t distribution becomes more peaked as n increases. In the limit as ν approaches infinity, the t distribution approaches the normal distribution. In fact, the t value from Appendix table 8 for infinite degrees of freedom is the same as the Z value. For small samples, however, the t values are considerably larger than the Z values from the normal distribution for the same level of α. When are the two sampling distributions sufficiently close so that the error from using the normal distribution to approximate the t is inconsequential? The answer that most statisticians have decided upon is when $n = 30$. Thus, with small samples ($n < 30$), we commonly use the t distribution as the basis for confidence intervals and tests on μ, but for large samples, we employ the normal approximation.

For *small samples* (n less than 30), we get the confidence interval for μ using the t distribution as follows. Let $t_{(\alpha/2,\ \nu)}$ be the value of the t distribution with $\nu = (n - 1)$ degrees of freedom that cuts off $\alpha/2$ of the probability in the upper tail, i.e., $P(t > t_{(\alpha/2,\ \nu)}) = \alpha/2$ and by symmetry $P(t < -t_{(\alpha/2,\ \nu)}) = \alpha/2$ for the lower tail. Then equation 6.9 becomes our statement of the

$$P(-t_{(\alpha/2,\ \nu)} < t < t_{(\alpha/2,\ \nu)}) = (1 - \alpha) \qquad \textbf{6.9}$$

confidence interval, and since t is defined in equation 6.8, we substitute this for t in the confidence interval probability statement in equation 6.9 and solve for μ to obtain our interval estimate, and our $100(1 - \alpha)$ percent confidence interval for μ appears in equation 6.10.

$$\bar{X} - t_{(\alpha/2,\ \nu)} \cdot S/\sqrt{n} < \mu < \bar{X} + t_{(\alpha/2,\ \nu)} \cdot S/\sqrt{n}. \qquad \textbf{6.10}$$

To illustrate the use of formula 6.10, suppose an animal scientist wants to estimate the average 205-day weaning weights for calves sired by a certain Gelbvieh bull. He takes a random sample of $n = 9$ calves from the herd known to be from the bull and obtains the following data:

550, 525, 570, 600, 485, 535, 580, 520, 540.

Construct a 95 percent confidence interval for the true average weaning weights. To do this, we first compute the values for the sample mean $\bar{X}$ and the sample standard deviation S; then we go to Appendix table 8 and obtain the appropriate values for $\pm t_{(\alpha/2,\ \nu)}$. Finally, we put all of the appropriate values in the expression for the confidence interval and obtain the result. We begin by putting the data in a column so that we can make the calculations (table 6.1)

$$\bar{X} = \Sigma X/n = 4{,}905/9 = 545$$
$$S^2 = \Sigma(X - \bar{X})^2/(n - 1) =$$
$$9{,}650/(9 - 1) = 9{,}650/8 = 1{,}206.25$$
$$S = \sqrt{S^2} = \sqrt{1{,}206.25} = 34.73$$

$$\nu = (n - 1) = (9 - 1) = 8$$
$$\alpha = 0.05, \text{ hence } t_{(\alpha/2,\ \nu)} = t_{(0.05/2,\ 8)} = t_{(0.025,\ 8)} = 2.306$$

We substitute the appropriate values into equation 6.10 to get:

$$\bar{X} - t_{(\alpha/2,\ \nu)} \cdot S/\sqrt{n} < \mu < \bar{X} + t_{(\alpha/2,\ \nu)} \cdot S/\sqrt{n}$$
$$545 - 2.306 \cdot 34.73/\sqrt{9} < \mu < 545 + 2.306 \cdot 34.73/\sqrt{9}$$
$$545 - 2.306\ (11.58) < \mu < 545 + 2.306\ (11.58)$$
$$545 - 26.7 < \mu < 545 + 26.7$$
$$518.3 < \mu < 571.7$$

The interval from 518.3 to 571.7 represents the 95 percent confidence interval for the average 205-day weaning weight for calves sired by this herd bull. Since it is quite large, we may want to select a larger sample. However, we do not get to work with large samples in all problems since we select some samples

TABLE 6.1 Calculation of mean and variance for weaning weight data

X (weights)	$X - \bar{X}$	$(X - \bar{X})^2$
550	5	25
525	−20	400
570	25	625
600	55	3,025
485	−60	3,600
535	−10	100
580	35	1,225
520	−25	625
540	−5	25
4,905	0	9,650

for conducting tests on the sample data and testing costs preclude taking large samples. In other cases, factors such as costs of obtaining the data, or the required use of experts, etc. make small samples the practical solution to the problem of estimating parameters.

For *large samples* ($n \geq 30$), we use the normal approximation to the t distribution as the basis for constructing 100 $(1 - \alpha)$ percent confidence intervals for μ, or to state this another way, we use S as a proxy for σ in the confidence interval from the previous section. Thus, the interval is

$$P(-Z_{\alpha/2} < Z < Z_{\alpha/2}) = (1 - \alpha) \qquad \boxed{6.11}$$

given in equation 6.11 and since

$$Z = (\bar{X} - \mu)/(S/\sqrt{n}),$$

we get the confidence interval in equation 6.12. Suppose we decide to increase the size of the

$$\bar{X} - Z_{\alpha/2} \cdot (S/\sqrt{n}) < \mu < \bar{X} + Z_{\alpha/2} \cdot (S/\sqrt{n}). \qquad \boxed{6.12}$$

sample in the calf weaning weight example from $n = 9$ to $n = 36$ animals. We collect weaning weights for a random sample of thirty-six calves and calculate $\bar{X} = 548$ pounds, $S = 20.5$ pounds from the sample data. We use equation 6.12 to obtain a 95 percent confidence interval on μ:

$$\bar{X} - Z_{\alpha/2} \cdot (S/\sqrt{n}) < \mu < \bar{X} + Z_{\alpha/2} \cdot (S/\sqrt{n})$$
$$548 - 1.96 \cdot (20.5/\sqrt{36}) < \mu < 548 + 1.96 \cdot (20.5/\sqrt{36})$$
$$548 - 1.96\,(3.42) < \mu < 548 + 1.96\,(3.42)$$
$$548 - 6.7 < \mu < 548 + 6.7$$
$$541.3 < \mu < 554.7$$

This interval is only 13.4 pounds wide, which is considerably less than the width of the interval we constructed for the small sample. It is smaller because several things occur. First, the large sample lets us use $Z_{\alpha/2}$ table values, which are somewhat less than the corresponding *t* values. Second, with the larger sample, we get a lower value for *S*, which usually occurs. And finally, the larger sample provides a lower value for $S_{\bar{X}}$ because we divide *S* by the square root of 36 rather than by the square root of 9.

Confidence Intervals for *p* from the Binomial Distribution

The binomial random variable *r* represents the number of successes in *n* trials of the experiment, and we know from an earlier chapter that the mean and variance of the binomial are

$$E(r) = np \text{ and } Var\ (r) = npq, \quad \textbf{6.13}$$

as stated in equation 6.13. Sometimes we are interested in another variable, $\hat{p}$, which is the proportion of successes in *n* trials when *n* represents a sample selected from a binomial population. We obtain $\hat{p}$ by dividing *r* by *n*, as in equation 6.14, which is a simple change of

$$\hat{p} = r/n \text{ or } \hat{p} = (1/n) \cdot r \quad \textbf{6.14}$$

variable. Because $(1/n)$ is a constant, the

$$E(\hat{p}) = (1/n) \cdot E(r) = (1/n) \cdot np = p.$$

Also,

$$Var(\hat{p}) = (1/n)^2 \cdot Var(r) = (1/n)^2 \cdot npq = pq/n.$$

Thus, $\hat{p}$ is an estimator of the population proportion of successes, *p*. We can easily construct a 100 $(1 - \alpha)$ percent confidence interval for *p* provided we use the normal approximation to the binomial. We can do that by setting the mean of the binomial equal to the mean of the normal and the standard deviation of the binomial equal to the standard deviation of the normal and using the *Z* formula. Thus $E(\hat{p}) = p$, and the standard deviation of $\hat{p} = \sqrt{pq/n}$. Unfortunately, we cannot use this last term in the *Z* formula because it contains the unknown parameter *p* that we are trying to estimate. So instead we use $\hat{p}$ in place of *p* and $(1 - \hat{p})$ in place of *q* to give our estimated standard deviation to be $\sqrt{\hat{p}(1 - \hat{p})/n}$ and we write *Z* as shown in equation 6.15.

$$Z = (\hat{p} - p)/\sqrt{[(\hat{p})(1 - \hat{p})/n]} \quad \textbf{6.15}$$

By using equation 6.11 and substituting the expression for *Z* from equation 6.15 into it and rearranging terms, we get the desired confidence interval on *p* as presented in equation 6.16.

$$\hat{p} - Z_{\alpha/2} \cdot \sqrt{[(\hat{p})(1 - \hat{p})/n]} < p < \hat{p} + Z_{\alpha/2} \cdot \sqrt{[(\hat{p})(1 - \hat{p})/n]}. \quad \textbf{6.16}$$

As an example, consider an animal scientist who takes a random sample of 100 cows from a large herd that have been artificially inseminated. Through testing, she determines that fifty-nine are pregnant. Construct a 95 percent confidence interval on p, the proportion of cows in the herd that are pregnant. To do this interval, we look up the $Z_{\alpha/2}$ value from Appendix table 7 and obtain $Z_{0.025} = 1.96$; calculate $\hat{p} = r/n = 59/100 = 0.59$, and then substitute these values into equation 6.16.

$$\hat{p} - Z_{\alpha/2} \cdot \sqrt{[(\hat{p})(1 - \hat{p})/n]} < p < \hat{p} + Z_{\alpha/2} \cdot \sqrt{[(\hat{p})(1 - \hat{p})/n]}$$

$$0.59 - 1.96 \cdot \sqrt{[(0.59)(1 - 0.59)/100]} < p < 0.59 + 1.96 \cdot \sqrt{[(0.59)(1 - 0.59)/100]}$$

$$0.59 - 1.96 \cdot \sqrt{0.002419} < p < 0.59 + 1.96 \cdot \sqrt{0.002419}$$

$$0.59 - 1.96 \cdot 0.049 < p < 0.59 + 1.96 \cdot 0.049$$

$$0.59 - 0.096 < p < 0.59 + 0.096$$

$$0.494 < p < 0.686$$

Thus, we estimate the true proportion of pregnant cows to be in the range of 0.49 to 0.69 for the herd after one exposure to artificial insemination with 95 percent confidence.

Confidence Interval for σ^2

In certain cases, we wish to construct an estimate for an unknown population variance. For instance, scientific instruments must provide unbiased readings with a very small error of measurement. A soil spectrometer that provides average readings that are correct is of little use to us if the variability in those readings is so high that the consistency from sample to sample for the same soils is missing. In farm machinery manufacturing, not only must the average size of parts produced be correct, but the variability in their size must be closely controlled or there is no standardization and parts are not interchangeable. Agricultural economists are concerned with price variability when constructing marketing plans for fruits and vegetables for producer groups. The sample variance, as presented in equation 6.17, is an unbiased estimator of σ^2 and the

$$S^2 = \frac{\Sigma (X - \overline{X})^2}{n - 1} \qquad \boxed{6.17}$$

distribution of sample variances obtained through repeated sampling belongs to the chi-square family of pdfs. This has been worked out by standardizing S^2 using a procedure similar to that for the Z variable. The standardized random variable χ^2, as defined in equation 6.18, belongs to the chi-square distribution if the population random variable X from which S^2 is calculated is

$$\chi^2 = \frac{(n - 1)S^2}{\sigma^2} \qquad \boxed{6.18}$$

normally distributed. Appendix table 9 contains probabilities in the tails of the chi-square distribution. Unlike the sampling distribution of $\overline{X}$, the distribution

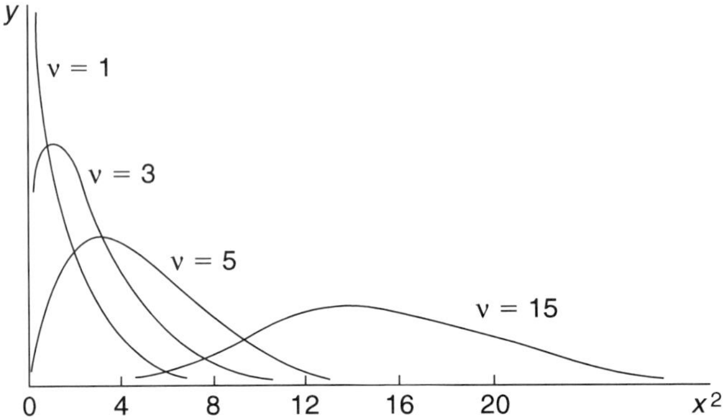

FIGURE 6.3 Chi-square distribution for selected values of nu

of S^2 is not symmetrical since S^2 cannot be negative. Hence, we have separate probabilities for each tail of the chi-square distribution. We denote the values of the right-tail probabilities by χ^2_α and the values of the left-tail probabilities by $\chi^2_{1-\alpha}$ for this distribution.

The shape of the chi-square distribution, like the t, depends upon ν, degrees of freedom associated with S^2. Thus, if we change the size of the sample, ν and the relevant chi-square pdf also change (figure 6.3). As the size of the sample increases, the form of the chi-square distribution approaches normal distribution. Thus, when degrees of freedom are large, for example 100, the resulting chi-square looks much more like the normal than when degrees of freedom are small, say 6.

To construct a 100 $(1 - \alpha)$ percent confidence interval for σ^2 when sampling from a normal population, we set up the expression in equation 6.19. Then we use the expression for chi-square

$$P(\chi^2_{(1-\alpha/2,\, \nu)} < \chi^2 < \chi^2_{(\alpha/2,\, \nu)}) = (1 - \alpha). \tag{6.19}$$

from equation 6.18 and substitute it into our probability statement, equation 6.20, rearrange terms, and we have the confidence interval we are seeking (equation 6.21).

$$\chi^2 = \frac{(n - 1)S^2}{\sigma^2}$$

$$\chi^2_{(1-\alpha/2,\, \nu)} < (n - 1)S^2/\sigma^2 < \chi^2_{(\alpha/2,\, \nu)} \tag{6.20}$$

$$\frac{(n - 1)S^2}{\chi^2_{(\alpha/2,\, \nu)}} < \sigma^2 < \frac{(n - 1)S^2}{\chi^2_{(1-\alpha/2,\, \nu)}} \tag{6.21}$$

In determining equation 6.21, we changed the sense of the inequality so that the larger value of chi-square from the table is now part of the term on the left, while the smaller table value is now a member of the term on the right.

Consider the following example confidence interval problem. A dairy co-operative is in the market for a new milk-packaging machine for half-gallon containers. A salesperson for the Sur-Fil Company makes a presentation for the product, claiming a range of no more than 0.4 ounce. The co-op manager obtains the location of such a machine from the salesperson and arranges to take a random sample from the production line. The manager obtains eight half-gallon containers off the line and computes their mean and variance as 64.1 and 0.018, respectively. Construct a 90 percent confidence interval for σ^2.

In solving this problem, we first compute degrees of freedom, $\nu = (n - 1) = (8 - 1) = 7$. From Appendix table 9, the values of $\chi^2_{0.05}$ and $\chi^2_{0.95}$ with 7 degrees of freedom are 14.07 and 2.17, respectively. Now with equation 6.21 we can compute the 90 percent confidence interval as:

$$\frac{(n - 1)S^2}{\chi^2_{(\alpha/2,\, \nu)}} < \sigma^2 < \frac{(n - 1)S^2}{\chi^2_{(1-\alpha/2,\, \nu)}}$$

$$(7)(0.018)/14.07 < \sigma^2 < (7)(0.018)/2.17$$

$$0.009 < \sigma^2 < 0.058$$

In interpreting this problem, the variance seems small, but if we take the variance at the upper limit of the interval, and take its square root to get the standard deviation, it is 0.24 ounces. If we construct a range of only twice the standard deviation, the range claimed by the salesperson seems questionable. Before making a purchasing decision, the manager may want to obtain more information.

Exercises

1. A simple random sample of 400 cotton farms revealed an average acreage of 615 with a standard deviation of 100. Construct a 90 percent confidence interval for the population mean. Interpret the confidence interval.
2. In a simple random sample of 1,000 sheep and goat raisers, 600 were in favor of new wool and mohair grading standards and 400 were opposed. Construct a 95 percent confidence interval for the true proportion of all sheep and goat raisers in favor of the new standards.
3. A farm machinery manufacturer ordered a large shipment of cultivator bolts. The quality control representative for the manufacturer took a random sample of 300 of the bolts and found thirty-six defectives. Construct a 99 percent confidence interval for the proportion of bolts in the shipment that is defective.
4. The mean scrotal circumference obtained from 100 randomly selected certificates for yearling bulls from a beef breed association was 31.5 cm with a standard deviation of 10 cm. Construct a 95.5 percent confidence interval on the average scrotal circumference for yearling bulls of this breed.
5. A random sample of nine yearling beef bulls of a particular breed was selected for semen evaluation. The average percent primary spermatozoal abnormalities was found to be 18 with a standard deviation of 2.7. Construct a 95 percent confidence interval on the true average percent of primary abnormalities for this breed.
6. A random sample of twenty-five cotton plants of a particular variety was selected for evaluation in a fruiting study. The average number of mature bolls per plant was twenty with a standard deviation of 10. Construct a 99 percent confidence interval for the true average number of mature bolls per plant for this variety.
7. In a germination study for exotic plants, it is important that the relative humidity in the greenhouse be kept within close tolerances or the plants fail to germinate. The horticulturist takes fifteen readings and finds the variance to be 0.6 percent relative humidity. Construct a 99 percent confidence interval on the actual variability in the humidity at the greenhouse.
8. Six judges score several pens of livestock on body conformation at a recent show. The variance in their ratings is found to be 10. Construct a 95 percent confidence interval on the true variability in the ratings of these judges.

CHAPTER

Statistical Estimation: Two Samples

If we want to compare two products or methods of performing a task or service, we must collect data from two samples and deal with two random variables. Unfortunately, there is not a well-known probability distribution we can use in estimating two random variables. However, we can apply a simple change of variable technique and use the now-familiar probability distributions from Chapter 6. For example, instead of having the two random variables $\bar{X}_1$ and $\bar{X}_2$, and using them separately to estimate the two population means, we form a unique random variable by subtracting them. Thus, the new random variable $(\bar{X}_1 - \bar{X}_2)$ belongs to a single-variable normal probability distribution with mean $(\mu_1 - \mu_2)$ and variance $(\sigma_1^2/n_1 + \sigma_2^2/n_2)$. The only part of this that may look strange is the variance. Variances cannot be negative, thus when we subtract random variables, we add the variances. Since we obtain point estimates by computing the values of the random variables from each sample, they are not difficult. However, we present confidence intervals in the next section for several random variables, each with different sampling distributions.

Confidence Intervals for $(\mu_1 - \mu_2)$ with σ_1, σ_2 Known

Consider the example in which a producer might wish to compare the average number of days to estrus for two different prostaglandin products for cattle. The producer takes a random sample from the first population, which has mean μ_1 and standard deviation σ_1 associated with animals treated with the first drug, and obtains a sample mean $\bar{X}_1$. Also, the producer takes a second independent random sample from the second population with mean μ_2 and standard deviation σ_2 associated with animals given the second drug, and computes the sample mean $\bar{X}_2$. If the producer draws repeated pairs of samples of sizes n_1 and n_2 from the two populations and calculates the estimate $(\bar{X}_1 - \bar{X}_2)$ for each pair, the result is a sampling distribution that is normal with mean equal to the difference between the population means and variance equal to the sum of the population variances. Thus, to compute a 100 $(1 - \alpha)$ percent confidence interval on the difference between the population means $(\mu_1 - \mu_2)$, we begin with the probability statement in equation 7.1 and substitute the expression

$$P(-Z_{\alpha/2} < Z < Z_{\alpha/2}) = (1 - \alpha) \tag{7.1}$$

in equation 7.2 for Z and rearrange terms to provide the statement we are seeking (equation 7.3).

$$Z = \frac{(\bar{X}_1 - \bar{X}_2) - (\mu_1 - \mu_2)}{\sqrt{\frac{\sigma^2_1}{n_1} + \frac{\sigma^2_2}{n_2}}} \tag{7.2}$$

We call the denominator of the formula in equation 7.2 the standard error of the difference between the means and its symbol is $\sigma(\bar{x}_1 - \bar{x}_2)$. This is the term that appears in the confidence interval statement in equation 7.3, which represents a 100 $(1 - \alpha)$ percent confidence interval on

$$(\bar{X}_1 - \bar{X}_2) - Z_{\alpha/2} \cdot \sigma(\bar{x}_1 - \bar{x}_2) < (\mu_1 - \mu_2) < (\bar{X}_1 - \bar{X}_2) + Z_{\alpha/2} \cdot (\sigma_{\bar{X}_1 - \bar{X}_2}) \tag{7.3}$$

the difference between two population means when we know the population standard deviations.

For an example of this process, consider a viticulturist who knows from laboratory tests the standard deviation of the sugar content of a red wine grape when it is produced on sandy soils to be $\sigma_1 = 5$ and when it is produced on sandy loam soils to be $\sigma_2 = 7.75$. However, she wants to estimate the average difference in sugar content for these soils under commercial growing conditions. Thus, she visits local vineyards and takes a random sample of size $n_1 = 25$ for this grape produced on sandy soils and obtains a sample mean of $\bar{X}_1 = 21$ and a second random sample for sandy loam soils of $n_2 = 30$ and computes $\bar{X}_2 = 26$. Compute a 98 percent confidence interval on the mean difference in sugar content for the red wine grape grown commercially on these two soils.

To write the confidence interval, we first obtain the $Z_{\alpha/2}$ value from Appendix table 7, which is ± 2.33. We next compute the standard error of the difference in the means as

$$\sigma_{(\bar{X}_1 - \bar{X}_2)} = \sqrt{(\sigma_1{}^2/n_1 + \sigma_2{}^2/n_2)} = \sqrt{(5^2/25 + 7.75^2/30)} = \sqrt{3.002} = 1.73.$$

Then we place all of the values in equation 7.3 and obtain the 98 percent interval

$$\begin{aligned}
(\bar{X}_1 - \bar{X}_2) - Z_{\alpha/2} \cdot \sigma_{(\bar{X}_1 - \bar{X}_2)} &< (\mu_1 - \mu_2) < (\bar{X}_1 - \bar{X}_2) + Z_{\alpha/2} \cdot \sigma_{(\bar{X}_1 - \bar{X}_2)} \\
(21 - 26) - 2.33 \cdot 1.73 &< (\mu_1 - \mu_2) < (21 - 26) + 2.33 \cdot 1.73 \\
-5 - 4.03 &< (\mu_1 - \mu_2) < -5 + 4.03 \\
-9.03 &< (\mu_1 - \mu_2) < -0.97.
\end{aligned}$$

Thus, we state with 98 percent confidence that the mean sugar content for the red wine grape produced commercially on sandy loam soils is between 0.97 and 9.03 units greater than when it is produced on sandy soils.

Confidence Intervals for $(\mu_1 - \mu_2)$ with σ_1, σ_2 Unknown

When we estimate the difference between two population means and do not know the population standard deviations, the relevant sampling distribution is the t distribution, but whether we actually use it depends upon the size of the samples. When both samples are large, i.e., $n_1, n_2 > 30$, we may use the sample variances as estimates for the respective population variances, but we cannot do this for small samples.

Large Samples

When we want estimates for the difference between two means and have access to data from two large samples, we use the normal distribution to approximate the t and the 100 $(1 - \alpha)$ percent confidence interval closely resembles the one from the previous section except that we calculate Z using the formula in equation 7.4. When we substitute this equation for Z in the

$$Z = \frac{(\bar{X}_1 - \bar{X}_2) - (\mu_1 - \mu_2)}{\sqrt{\frac{S^2_1}{n_1} = \frac{S^2_2}{n_2}}} \quad \boxed{7.4}$$

expression

$$P(-Z_{\alpha/2} < Z < Z_{\alpha/2}) = (1 - \alpha)$$

and manipulate it algebraically, we obtain the desired confidence interval in equation 7.5, in which

$$(\bar{X}_1 - \bar{X}_2) - Z_{\alpha/2} \cdot S_{(\bar{X}_1 - \bar{X}_2)} < (\mu_1 - \mu_2) < (\bar{X}_1 - \bar{X}_2) + Z_{\alpha/2} \cdot S_{(\bar{X}_1 - \bar{X}_2)} \quad \boxed{7.5}$$

$S_{(\bar{X}_1 - \bar{X}_2)} = \sqrt{(S_1^2/n_1 + S_2^2/n_2)}$ is an estimate of the population standard error of the difference between the means based on the sample variances.

Consider the following example. A cultivator part can be manufactured by an extrusion process or it can be fabricated with a stamping machine and welded together. The project manager is concerned about the strength of the part made by the two processes. He has several parts made by each process and computes the statistics (table 7.1). Construct a 99 percent confidence interval on the average difference in the strength of the part made by fabrication and stamping.

To construct the confidence interval, we first obtain the $Z_{\alpha/2}$ values from Appendix table 7. For a 99 percent confidence interval, $\alpha = 0.01$ and the appropriate table values are ± 2.57. Next we compute

$$S_{(\bar{X}_1 - \bar{X}_2)} = \sqrt{(S_1^2/n_1 + S_2^2/n_2)} = \sqrt{(256/64 + 171.5/49)} = \sqrt{7.5} = 2.74.$$

TABLE 7.1 Sample Data Obtained for Fabricating and Stamping and Welding Cultivator Parts

Item	Fabrication Process	Stamping and Welding Process
Sample size	64	49
Average strength, foot pounds	510	490
Sample variance	256	171.5

Then we place the appropriate values in equation 7.5 and solve

$$(\bar{X}_1 - \bar{X}_2) - Z_{\alpha/2} \cdot S_{\bar{X}_1 - \bar{X}_2} < (\mu_1 - \mu_2) < (\bar{X}_1 - \bar{X}_2) + Z_{\alpha/2} \cdot S_{\bar{X}_1 - \bar{X}_2}$$
$$(510 - 490) - 2.57 \cdot 2.74 < (\mu_1 - \mu_2) < (510 - 490) + 2.57 \cdot 2.74$$
$$20 - 7.04 < (\mu_1 - \mu_2) < 20 + 7.04$$
$$12.96 < (\mu_1 - \mu_2) < 27.04$$

According to the preceding 99 percent confidence interval, the cultivator part made with the extrusion process is between 12.96 and 27.04 foot pounds stronger on the average than the one made by stamping and welding.

Small, Independent Samples

When we have small samples and we draw them in a way so that they are independent, the *t* distribution provides the sampling distribution for the random variable—the difference between the population means. In this instance, small samples do not provide very stable estimates of the variance, and we use all of the data for estimating the variance, i.e., we *pool* the data according to the formula in equation 7.6. When we pool variances, we weigh them by degrees of freedom,

$$S^2 = \frac{(n_1 - 1)S_1^2 + (n_2 - 1)S_2^2}{n_1 + n_2 - 2} \quad \boxed{7.6}$$

add the result, and then divide by total degrees of freedom.[1] To obtain $S_{\bar{X}_1 - \bar{X}_2}$, we divide the pooled variance by each sample size, sum the result, and take the square root, as in equation 7.7.

$$S_{\bar{X}_1 - \bar{X}_2} = \sqrt{S^2/n_1 + S^2/n_2}. \quad \boxed{7.7}$$

To get a 100 (1 − α) percent confidence interval with the t distribution, we write

$$P(-t_{(\alpha/2,\, \nu)} < t < t_{(\alpha/2,\, \nu)}) = (1 - \alpha)$$

$$t = \frac{(\bar{X}_1 - \bar{X}_2) - (\mu_1 - \mu_2)}{\sqrt{\frac{S^2}{n_1} + \frac{S^2}{n_2}}} \quad \boxed{7.8}$$

where t is defined in equation 7.8. When we substitute the right side of equation 7.8 for t into the confidence interval statement and rearrange terms, we get the confidence interval on $(\mu_1 - \mu_2)$ we are seeking in equation 7.9.

$$(\bar{X}_1 - \bar{X}_2) - t_{(\alpha/2,\, \nu)} \cdot S_{\bar{X}_1 - \bar{X}_2} < (\mu_1 - \mu_2) < (\bar{X}_1 - \bar{X}_2) + t_{(\alpha/2,\, \nu)} \cdot S_{\bar{X}_1 - \bar{X}_2} \quad \boxed{7.9}$$

Consider the following example. A pork producer feeds one litter of ten pigs a vitamin supplement in their drinking water and uses a second litter of nine pigs as a control. The producer tests weight gain from birth to weaning and notes that the pigs fed the vitamin supplement gained an average of 25 pounds with a standard deviation of 8 pounds, while the control group gained an average of 20 pounds with a standard deviation of 11 pounds. Construct a 95 percent confidence interval on the difference in the mean weight gain for the two groups.

To construct the interval, we first compute total degrees of freedom as

$$\nu = (n_1 + n_2 - 2) = (10 + 9 - 2) = 17.$$

Since this is a 95 percent confidence interval, $\alpha = 0.05$ and, from Appendix table 8, the appropriate t values are ± 2.11. Next, we compute the pooled variance as

$$S^2 = [(n_1 - 1)S_1{}^2 + (n_2 - 1)S_2{}^2]/(n_1 + n_2 - 2)$$
$$S^2 = [(10 - 1)8^2 + (9 - 1)11^2]/(10 + 9 - 2)$$
$$S^2 = [576 + 968]/17 = 90.82$$

and use it to compute the standard error of the difference between the means,

$$S_{\bar{X}_1 - \bar{X}_2} = \sqrt{S^2/n_1 + S^2/n_2}$$
$$S_{\bar{X}_1 - \bar{X}_2} = \sqrt{90.82/10 + 90.82/9}$$
$$S_{\bar{X}_1 - \bar{X}_2} = \sqrt{9.082 + 10.091} = \sqrt{19.173}$$
$$S_{\bar{X}_1 - \bar{X}_2} = 4.38.$$

Finally, we write the expression for the confidence interval on the difference between the means (equation 7.9) and evaluate it using the values we have obtained:

$$(\bar{X}_1 - \bar{X}_2) - t_{(\alpha/2,\, \nu)} \cdot S_{\bar{X}_1 - \bar{X}_2} < (\mu_1 - \mu_2) < (\bar{X}_1 - \bar{X}_2) + t_{(\alpha/2,\, \nu)} \cdot S_{\bar{X}_1 - \bar{X}_2}$$
$$(25 - 20) - 2.11 \cdot 4.38 < (\mu_1 - \mu_2) < (25 - 20) + 2.11 \cdot 4.38$$
$$5 - 9.24 < (\mu_1 - \mu_2) < 5 + 9.24$$
$$-4.24 < (\mu_1 - \mu_2) < 14.24$$

This 95 percent confidence interval spans zero since it goes from -4.24 to 14.24. Thus, there could be no difference in the weaning weights of the pigs that were fed the vitamin supplement versus the control group since $(\mu_1 - \mu_2) = 0$ is contained in the confidence interval. On the basis of this experiment, the pork producer would not be willing to purchase a vitamin supplement for his or her pigs. This is a relatively wide confidence interval that sometimes occurs with small samples when the sample variances are large, like we see here. We may wish to take steps to decrease the variances and retry the experiment, if that is an option.

Small, Paired Samples

If we have small sample sizes and **paired data,** i.e., the data are not independent because some factor is present that makes them related, then we use the t distribution as the basis for a 100 $(1 - \alpha)$ percent confidence interval on $(\mu_1 - \mu_2)$, provided we construct the interval with data representing the differences in the observations for the two samples as opposed to the actual observations themselves. Paired data occur because of the type of experiment we are conducting. In this case, the experiment involves observations on the same subjects before and after we administer a treatment, or it contains observations on two samples that are affected by some variable that could be a block in a randomized block design, such as soil type in a cotton yield experiment. We require the two samples to be the same size so that we can pair the observations and obtain differences. The subtraction of one sample value from another essentially eliminates the influence of the factor that serves to make the samples dependent. For example, inheritance affects the ability of animals to gain weight. Thus, we prefer to use animals in weight-gain studies that are related, such as lamb twins, and conduct a paired t analysis on the results.

We derive the confidence interval statement from the now-familiar probability statement:

$$P(-t_{(\alpha/2,\, \nu)} < t < t_{(\alpha/2,\, \nu)} = (1 - \alpha),$$

but in this case the formula for t is different. We define the random variable d as the difference between the sample observations, i.e., $d_1 = X_{11} - X_{21}$, $d_2 = X_{12} - X_{22}, \ldots, d_n = X_{1n} - X_{2n}$ for all n pairs of data, $\bar{d}$ as the mean of differences, and S_d^2 as the variance of the differences. These latter statistics are computed just like those for any single sample, i.e., $\bar{d} = \Sigma d/n$, etc. Thus, the formula for t appears as in equation 7.10. By substituting the values on the right-hand side of the

$$t = \frac{\bar{d} - (\mu_1 - \mu_2)}{\dfrac{S_d}{\sqrt{n}}} \qquad \boxed{7.10}$$

equation for t in the expression for the confidence interval, and rearranging terms, we get the desired confidence interval for $(\mu_1 - \mu_2)$ shown in equation 7.11.

$$\bar{d} - t_{(\alpha/2,\, \nu)} \cdot S_d/\sqrt{n} < (\mu_1 - \mu_2) < \bar{d} + t_{(\alpha/2,\, \nu)} \cdot S_d/\sqrt{n} \qquad \boxed{7.11}$$

Consider the following example. To test two new grain sorghum hybrids under ordinary growing conditions, a seed company selects nine farms at random and has each farmer plant both hybrids in experimental plots. The yields in hundredweight (cwt.) per acre for the nine locations are given in table 7.2. Construct a 95 percent confidence interval for the mean difference in yield for the two hybrids.

To obtain the 95 percent confidence interval, we first calculate degrees of freedom as

$$\nu = (n - 1) = (9 - 1) = 8$$

TABLE 7.2 Grain sorghum yields in cwt. per acre for nine different farms

Farm	Hybrid 1	Hybrid 2
1	85	79
2	88	80
3	58	60
4	94	92
5	85	78
6	93	87
7	74	75
8	80	76
9	101	95

and obtain the t values for $\alpha = 0.05$ from Appendix table 8 as ± 2.306. Next, we compute the two statistics $\bar{d}$ and S_d by setting up the random variable d and making the necessary calculations by using the appropriate formulas for the mean and standard deviation for d (table 7.3).

$$\bar{d} = \Sigma d/n = 36/9 = 4$$

$$S_d^2 = \Sigma(d - \bar{d})^2/(n - 1) = 102/(9 - 1) = 12.75$$

$$S_d = \sqrt{S_d^2} = \sqrt{12.75} = 3.57.$$

TABLE 7.3 Calculations for the mean and variance of *d* for the hybrid sorghum yield data

Farm	Hybrid 1	Hybrid 2	d	$(d - \bar{d})$	$(d - \bar{d})^2$
1	85	79	6	$6 - 4 = 2$	4
2	88	80	8	$8 - 4 = 2$	16
3	58	60	−2	$-2 - 4 = -6$	36
4	94	92	2	$2 - 4 = -2$	4
5	85	78	7	$7 - 4 = 3$	9
6	93	87	6	$6 - 4 = 2$	4
7	74	75	−1	$-1 - 4 = -5$	25
8	80	76	4	$4 - 4 = 0$	0
9	101	95	6	$6 - 4 = 2$	4
Totals			36	0	102

By substituting the values we obtained thus far into the confidence interval (equation 7.11), we get the result:

$$\bar{d} - t_{(\alpha/2,\, \nu)} + S_d/\sqrt{n} < (\mu_1 - \mu_2) < \bar{d} + t_{(\alpha/2,\, \nu)} \cdot S_d/\sqrt{n}$$
$$4 - 2.306 \cdot 3.57/\sqrt{9} < (\mu_1 - \mu_2) < 4 + 2.306 \cdot 3.57/\sqrt{9}$$
$$4 - 2.74 < (\mu_1 - \mu_2) < 4 + 2.74$$
$$1.26 < (\mu_1 - \mu_2) < 6.74.$$

This 95 percent confidence interval states that the average yield of Hybrid 1 is between 1.26 and 6.74 cwt. higher than that of Hybrid 2. By pairing the data, we take care of individual differences from farm to farm that affect yield such as rainfall, inherent soil fertility, and management ability.

Confidence Intervals for $p_1 - p_2$

We can construct a 100 $(1 - \alpha)$ percent confidence interval on the difference between two sample proportions by using the normal approximation to the binomial as the sampling distribution with the random variable $(\hat{p}_1 - \hat{p}_2)$ as the estimator of the difference in the population proportions $(p_1 - p_2)$, where $\hat{p}_1$ is defined as r_1/n_1 and $\hat{p}_2 = r_2/n_2$. The confidence interval statement is

$$P(-Z_{\alpha/2} < Z < Z_{\alpha/2}) = (1 - \alpha)$$

where Z is defined as in equation 7.12. We obtain the confidence interval on $(p_1 - p_2)$ by substituting the

$$Z = \frac{(\hat{p}_1 - \hat{p}_2) - (p_1 - p_2)}{\sqrt{\dfrac{\hat{p}_1(1 - \hat{p}_1)}{n_1} + \dfrac{\hat{p}_2(1 - \hat{p}_2)}{n_2}}} = \frac{(\hat{p}_1 - \hat{p}_2) - (p_1 - p_2)}{S_{p_1 - p_2}} \qquad \mathbf{7.12}$$

right portion of equation 7.12 for Z into the original confidence interval probability statement and rearranging terms to give equation 7.13.

$$(\hat{p}_1 - \hat{p}_2) - Z_{\alpha/2} \cdot S_{p_1 - p_2} < (p_1 - p_2) < (\hat{p}_1 - \hat{p}_2) + Z_{\alpha/2} \cdot S_{p_1 - p_2} \qquad \mathbf{7.13}$$

Consider the following example. A horticulturist wishes to estimate the difference in the proportion of grape cuttings that root in the greenhouse. She treats one set of 160 cuttings with a root stimulant and observes 140 live plants at the end of the experiment, and leaves the other set of 120 cuttings untreated and obtains 96 live plants. Construct a 95 percent confidence interval on the difference in the proportion of live plants for this experiment.

To construct the confidence interval, we first find the appropriate Z value from Appendix table 7. When $\alpha = 0.05$, the table $Z_{\alpha/2} = \pm 1.96$. Next, we compute

$$\hat{p}_1 = r_1/n_1 = 140/160 = 0.875;$$
$$\hat{p}_2 = r_2/n_2 = 96/120 = 0.800; \qquad \text{and}$$
$$S_{p_1 - p_2} = \sqrt{\hat{p}_1(1 - \hat{p}_1)/n_1 + \hat{p}_2(1 - \hat{p}_2)/n_2}$$
$$S_{p_1 - p_2} = \sqrt{(0.875)(1 - 0.875)/160 + (0.800)(1 - 0.800)/120}$$

$$S_{p_1 - p_2} = \sqrt{0.1094/160 + 0.1600/120}$$
$$S_{p_1 - p_2} = \sqrt{0.00068 + 0.00133}$$
$$S_{p_1 - p_2} = \sqrt{0.002} = 0.04$$

Now we have sufficient information to construct the confidence interval using equation 7.13.

$$(\hat{p}_1 - \hat{p}_2) - Z_{\alpha/2} \cdot S_{p_1 - p_2} < (p_1 - p_2) < (\hat{p}_1 - \hat{p}_2) + Z_{\alpha/2} \cdot S_{p_1 - p_2}$$
$$(0.875 - 0.800) - 1.96 \cdot 0.04 < (p_1 - p_2) < (0.875 - 0.800) + 1.96 \cdot 0.04$$
$$0.075 - 0.078 < (p_1 - p_2) < 0.075 + 0.078$$
$$-0.003 < (p_1 - p_2) < 0.153$$

Since this 95 percent confidence interval just spans zero, the horticulturist may desire more information before using the root stimulant; her estimate is −0.003 to 0.153 as the difference in the proportion of grape cuttings that rooted with the root stimulant and those that rooted without.

Endnote

[1] We can illustrate the formula for total degrees of freedom as follows: $(n_1 - 1) + (n_2 - 1)$ can be rewritten as $(n_1 + n_2 - 2)$.

Exercises

1. In a simple random sample of 100 market hogs of a mixed breed, the average selling weight was 220 pounds with a standard deviation of 20 pounds, while for a sample of 400 pure breeds, the average selling weight was 210 pounds with a standard deviation of 30 pounds. Construct a 95 percent confidence interval on the true difference in average selling weight for the two groups of market hogs.
2. In trying to decide which brand of baling wire to purchase, a farmer obtains snippets of fifty pieces of Tie-Wright brand and sixty pieces of Strong Stuf brand. The farmer has them tested at a nearby lab for tensile strength. The lab reports the following:
   - Average tensile strength of Tie-Wright is 95.8 pounds with a standard deviation of 5 pounds.
   - Thirty-four percent of the Tie-Wright wire has a tensile strength of 100 pounds or more.
   - Average tensile strength of Strong Stuf is 99 pounds with a standard deviation of 6 pounds.
   - Thirty percent of Strong Stuf wire has a tensile strength of 100 pounds or more.

   a. Construct a 90 percent confidence interval for the average difference in tensile strength for the two brands of wire.
   b. Construct a 95 percent confidence interval for the actual difference in the proportion of wire that has a tensile strength of 100 pounds or more for the two brands.
   c. Interpret the results.
3. In evaluating the germination of mountain laurel seeds, 100 were scarified and planted and seventy-five were acid-treated and planted. Fifty of the scarified seeds germinated while forty-eight of the acid-treated seeds germinated. Construct a 95.5 percent confidence interval on the difference in the true proportion germinated by the two treatments.
4. In a random sample of ten melon farms, harvest laborers were paid an average of $7.25 per hour with a standard deviation of $1.25, while a sample of twelve fruit farms paid an average of $8 per hour with a standard deviation of $1. Construct a 90 percent confidence interval on the true average difference in hourly wages paid to harvest laborers by the two types of farms.
5. In a random sample of eight wheat growers, average selling costs were $0.10 per bushel with a standard deviation of $0.05, while a random sample of fifteen barley growers paid an average of $0.07 per bushel selling costs with a standard deviation of $0.03. Construct a 95 percent confidence interval on the true average difference in selling costs per bushel.
6. Six sets of twin dairy heifers with their first calves by their sides were fed identical rations except that one twin also got BST, a hormone supposed to increase milk production. Construct a 95 percent confidence interval on the difference in milk production (in pounds per animal).

Twins	Pounds of Milk, No BST	Pounds of Milk, BST
1	45	48
2	50	51
3	60	64
4	55	56
5	53	57
6	49	52

7. The following data represent paired yields of two varieties of barley. Each pair was planted on a different farm. Estimate the average difference in yields with a 95 percent confidence interval.

Pair	Hi-Yield	Ocala 95
1	37	42
2	41	45
3	35	33
4	52	56
5	44	50
6	30	32
7	38	35
8	54	55

CHAPTER

Hypothesis Testing

The procedures presented in the preceding chapter concern the process for making both point and interval estimates of population parameters, with the interval estimates providing an expression of the degree of uncertainty about the true value of the parameters. Intervals also allow us to test certain assumed values of the population parameters. We are interested in such values, which are called statistical hypotheses. A **statistical hypothesis** is defined as an assumption we make about some parameter or population variable.

We cannot classify all hypotheses as statistical. For example, the statement that "Braunvieh cattle are the best beef breed in the United States" is not statistical. However, with slight modification, we can make it into a statistical hypothesis. The statement that "Braunvieh is the breed most favored by beef producers in the United States" is statistical since it presents an assumption about a parameter—the proportion of beef breeders in the United States that we can test with data.

Types of Tests

To establish whether a statistical hypothesis is true with complete certainty may require us to test the entire population. Because this is generally impractical for a number of reasons, we almost always take a sample and test a statistical hypothesis with the sample data. We already know from previous discussions that sampling introduces possible error. Thus, in testing a statistical hypothesis, we set the amount of error that we are willing to allow, and generally make it very small. In addition, tests vary in complexity depending upon whether the statistical hypothesis is simple or composite. A simple hypothesis provides only one value of the parameter to test, e.g., H_o: $\mu = 75$, is a simple hypothesis about the population mean. We commonly use the notation H_o: to denote that a hypothesis is being proposed for testing. The statement after the colon is the actual hypothesis. In contrast, a composite hypothesis provides a range of values of the parameter for testing. The range may contain only a small number or a nearly infinite set of values. In testing a composite hypothesis, we must decide whether the parameter takes on any one of the possible values in the range. Examples of composite hypotheses include H_o: $\mu < 150$ and H_o: $\mu_1 - \mu_2 > 0$.

When we are setting up a test of statistical hypotheses, we specify two statements and decide in favor of one over the other. These are called the **null** and **alternative hypotheses.** The null hypothesis, denoted H_o, is the statistical statement of no difference, or no effect, or nothing. The alternative hypothesis, H_a or H_1 (we commonly see both notations), is the statistical statement of the population parameter values that we believe are true. Thus, in a test of hypothesis, we want to reject the null hypothesis (thereby claiming it is false and accepting the alternative as true). In tests in which we cannot reject the null hypothesis, we do not say it is true, we just do not reject it because in some future test, the hypothesis may be rejected. In scientific inquiry, there is always the possibility that new data will cause us to reject as false a hypothesis that we had come to believe was true.

The null and alternative hypotheses can both be simple or composite or some combination. For example, we may test the simple null hypothesis H_o: $p = 0.166$ against a simple alternative H_a: $p = 0.25$, or against a composite alternative H_a: $p \neq 0.166$. Or we could test the composite null hypothesis H_o: $\mu < 10$ against a simple alternative such as H_1: $\mu = 10$ or against a composite alternative H_1: $\mu \geq 10$. Whatever form the hypotheses may take, the true value of the parameter must be either in the set specified by H_o or in the set defined by H_a, i.e., the null and alternative hypotheses must account for all of the appropriate values the parameter may assume. One means of doing this is to make the two sets complementary, e.g., if we have H_o: $\mu > 50$, then the complementary hypothesis is H_a: $\mu \leq 50$. The hypotheses in the preceding example lead to a one-sided test because if the null hypothesis is not true, all of the values of the alternative are on one side of it (the left in this case). Thus, there are one-sided tests to the left and one-sided tests to the right, depending upon whether the alternative hypothesis specifies values less than or greater than those in the null. In the case in which we use a simple null hypothesis with a composite alternative, and variation in the parameter values can be on both sides of the null value, then we have a two-sided test. For example, the statement H_o: $p = 0.125$ and the accompanying alternative H_a: $p \neq 0.125$, implies a two-sided test.

In tests of hypotheses, our problem is how to use sample data to make a choice between two mutually exclusive statements about the value of the parameter. As we stated earlier, such a decision may involve error. Specifically, there are two kinds of possible error: the decision to accept the alternative hypothesis when the null is actually true (a **Type I error**), or the decision to accept the null hypothesis when the alternative is true (a **Type II error**).

Type I and Type II Errors

To illustrate Type I and Type II errors, suppose, for example, that an agricultural economist wishes to test average annual household spending for fresh fruits and vegetables. The economist sets up the null hypothesis H_o: $\mu \leq \$1,200$ versus

TABLE 8.1 Type I and Type II Errors

Decision	State of the World	
	H_o True	H_a True
Accept H_o	Correct decision	Type II error with probability β
Reject H_o	Type I error with probability α	Correct decision

the alternative H_a: $\mu > \$1,200$. She commits a Type I error if the test results lead her to accept H_a when spending is actually $1,200 or less, or a Type II error if she does not reject H_o when spending is greater than $1,200.

Since it is possible to make a Type I or Type II error in most tests of statistical hypotheses, we want to design a procedure to definitely take these errors into account. To do this, we must be concerned with their probability of occurrence. In the preceding chapter, the Greek letter α indicated the probability that a confidence interval might not contain the true value of the population parameter and we set it to a small value. Now, we use α to represent the small probability of making a Type I error and, again, we set it. We denote the probability of making a Type II error by the Greek letter beta (β). Developments leading to these errors are stated in table 8.1. To illustrate the calculation of α and β, we consider the following. Suppose we test the "fairness" of a die. We are told that a die weighted so that the proportion of 1s that turn up is $p = 0.25$, and the die has been inadvertently mixed with several fair dice in our possession. It has no identifying marks, so we propose the following test. We will randomly select a die and toss it $n = 100$ times and observe the number of 1s that turn up, say r. We compute the proportion of ones as stated in equation 8.1,

$$\hat{p} = r/n \tag{8.1}$$

and if $\hat{p}$ is no larger than some reasonable value, we will accept the die being tested as fair, but if $\hat{p}$ is larger than our value, we will reject the hypothesis that the die is fair and will have located the weighted die. Thus we have a decision rule to govern our actions. The reasonable value that we select is called the **critical value.** Suppose we set the critical value as $p = 0.20$, i.e., if no more than twenty 1s occur in 100 tosses of the die, we do not reject the null hypothesis and call the die fair, but if more than twenty 1s occur, we reject the null hypothesis in favor of the alternative and call the die unfair, hence weighted. Thus, we have the following hypotheses:

$$H_o\text{: } p = 0.167$$
$$H_a\text{: } p = 0.250$$

and the decision rule is $p = 0.20$. We illustrate this as follows with the normal approximation to the binomial, which gives a mean of $\mu = p$ and a standard deviation of $\sigma = \sqrt{pq/n}$. There are actually two sampling distributions—one when the null hypothesis is true ($\mu = 0.167$, $\sigma = 0.037$) and another when the alternative hypothesis is true ($\mu = 0.25$, $\sigma = 0.043$). The α represents the vertically

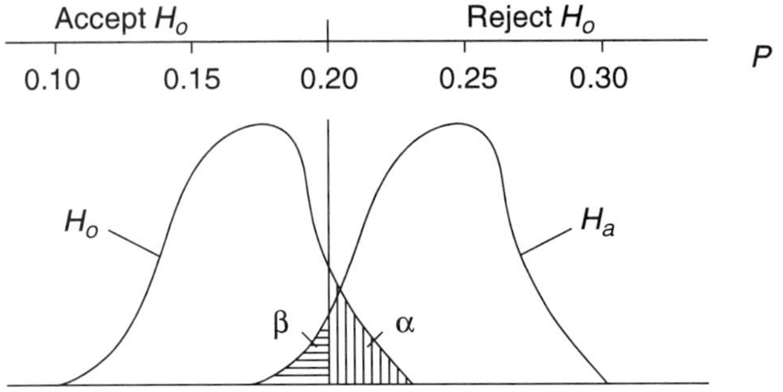

FIGURE 8.1 Sampling distributions for the null and alternative hypotheses for the die experiment

striped area under the first distribution that is to the right of the decision rule ($p = 0.20$), while β represents the horizontally striped area under the second distribution that lies to the left of the decision rule (figure 8.1). Note that we must specify the decision rule before we conduct the test and also before we can know the values of α and β. We now use the Z formula to calculate the actual probabilities of the Type I and Type II errors. We recall that we can calculate Z as in equation 8.2,

$$Z = (\hat{p} - p)/\sqrt{pq/n} \qquad \boxed{8.2}$$

and for this problem, we have set $\hat{p} = 0.20$ as the value of the decision rule. To calculate Z_α, we use the values associated with the sampling distribution when H_o is true, i.e., $p = 0.167$ and $\sigma_p = 0.037$. Thus

$$Z_\alpha = (0.20 - 0.167)/0.037 = 0.89;$$

and the probability of a Z exceeding 0.89 from Appendix table 7 is equal to

$$0.5000 - 0.3133 = 0.1867 = \alpha$$

for this test. We can also compute the probability of a Type II error by the Z formula with values from the sampling distribution when H_a is true. In this instance,

$$Z_\beta = (0.20 - 0.25)/0.043 = -1.16.$$

From Appendix table 7, we find the probability of a Z less than -1.16 by subtracting the table probability from 0.5, i.e.,

$$0.5000 - 0.3770 = 0.1230 = \beta$$

for this test.

Since the critical value in the decision rule is ultimately what determines the sizes of the Type I and Type II errors once we specify the null and alternative hypotheses, we can move the decision rule out, further away from the value of

H_o: $p = 0.167$ to $p = 0.22$, and thus reduce the probability of a Type I error if we think that $\alpha = 0.1867$ is too large for this test. But then we increase the probability of β since the decision rule line affects both errors, and there is a trade-off involved—one error probability cannot be reduced without increasing the other. We will return to this later.

But for now, what is the interpretation of a Type I error in the context of our test concerning the fairness of the die? We toss a die 100 times and the value of $\hat{p}$ that we compute from our experiment is greater than 0.20. Thus, on the basis of our decision rule, we reject the null hypothesis. According to the result of this test, we get more than 20 percent 1s and thus conclude that the die is not fair. A Type I error occurs if, in fact, the die is fair. Somehow our experiment led to a $\hat{p}$ that is relatively high even though the die is fair. This happens 100 α percent of the time with the way we constructed this test, or 18.67 percent of the time. Thus, in this case, we commit a Type I error by calling a fair die crooked.

Similarly, a Type II error can occur. In this case, we toss the die 100 times and compute a $\hat{p}$ value that is small, less than 0.20, and on the basis of the decision rule, we fail to reject the null hypothesis, i.e., we conclude the die is fair. A Type II error occurs if, in fact, the die is crooked. It is actually weighted so that, on average, the proportion of 1s that turn up is $p = 0.25$, but in this instance, we get some small proportion of 1s with the die. Our test calls it fair and this happens with probability β, which is 0.1230 or in 12.3 percent of the time. The measure $1 - \beta = 1 - 0.1230 = .8770$ is called the **power of the test.**

Controlling Both α and β

Since both of these errors are undesirable and their probabilities can be large, let's see how to choose a decision rule that balances them. In our die example, we first specified a critical value for the decision rule and then computed the corresponding values of α and β. Since these risks are related to the sample size, and increasing n decreases both α and β, our choice of a critical value also depends on the sample size. Thus, in practice, we select the size of α, which determines the critical value, and the sample size, and then compute β for values of H_a that are important. If β is too large, then we compute the sample size, which provides an acceptable value, and increase it to that size, if possible. Notice that we must do such calculations in the planning stage of a test of hypotheses or they are of little value.

Tests of Hypotheses

Now we are ready to discuss some of the more standard tests of hypotheses for one and two populations. Tests involving three or more populations will be discussed later. We begin with the most common tests of hypotheses, which are one-sample tests that involve μ, the population mean.

The following steps are involved in a test of hypotheses. We follow these in succeeding tests even though we may not stop to identify them. The steps are:

1. State the null and alternative hypotheses.
2. Determine the sampling distribution for the test, such as Z, t, etc.
3. State α and determine the critical value for the test and the decision rule.
4. Compute the value of the test statistic using available sample data.
5. Make the appropriate decision.

One-Sample Tests for μ with σ Known

If we know the population standard deviation, we may test the null hypothesis, H_o: $\mu = \mu_o$ against an alternative that is either one-sided (H_a: $\mu > \mu_o$ or H_a: $\mu < \mu_o$) or two-sided (H_a: $\mu \neq \mu_o$). The sampling distribution for such tests is the normal distribution and the test statistic is a computed Z value, as given in equation 8.3 We compare the computed Z to a Z from Appendix

$$Z = \frac{\overline{X} - \mu_o}{\frac{\sigma}{\sqrt{n}}} \tag{8.3}$$

table 7, which cuts off a probability of α in the appropriate tail of the normal sampling distribution for $\overline{X}$. How do we know which is the appropriate tail? We look at the alternative hypothesis statement. If it says that $\mu < \mu_o$, then we have a one-sided test to the left and the left tail of the sampling distribution contains α, the computed Z is negative, and the Z from Appendix table 7 is also negative. Alternatively, if we construct H_a so that $\mu > \mu_o$, we have a one-sided test to the right, the critical region is in the right tail of the sampling distribution, and both the computed and table Zs are positive. Or, we could have a two-tailed test resulting from an alternative hypothesis such as $\mu \neq \mu_o$, which requires that we split α and put half in each tail of the sampling distribution so that there are two critical regions and two table Z values. Computed Z may be either positive or negative, but we compare it to the table Z values to make a decision to reject or not reject the null hypothesis.

Let's consider two examples—one for a one-tailed test and another for a two-tailed test—to help clarify the preceding discussion. First, consider a one-tailed test. Suppose we know that the daily wages of farm workers are normally distributed with mean, $\mu_o = \$55$ and standard deviation, $\sigma = \$10$. Farmers in a particular region are accused of paying inferior wages. We take a random sample of $n = 36$ from the area and compute $\overline{X} = \$52$. If we select $\alpha = 0.05$, can farmers in this area be accused of paying inferior wages?

To solve this problem, we first write the hypotheses:

$$H_o: \mu = \$55$$
$$H_a: \mu < \$55$$

The alternative hypothesis implies a one-tailed test to the left. Since $\alpha = 0.05$, Z from Appendix table 7 is -1.64. Thus, the critical region is in the left tail of the sampling distribution and is defined by the critical value $Z_\alpha = -1.64$ (figure 8.2).

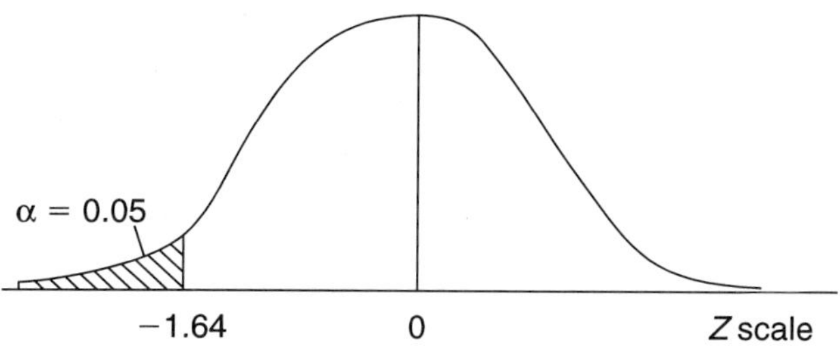

FIGURE 8.2 Sampling distribution and critical region for one-tailed test of μ

If the computed Z lies to the left of −1.64, in the critical region, we reject H_o and conclude that the farmers in the region do pay inferior wages. But if computed Z falls between 0 and −1.64, we cannot reject the null hypothesis. We compute Z from equation 8.3 as follows:

$$Z = \frac{\overline{X} - \mu_o}{\frac{\sigma}{\sqrt{n}}} = \frac{52 - 55}{\frac{10}{\sqrt{36}}} = -1.80$$

We make the decision in this test by comparing the computed Z value of −1.8 to the table value of −1.64. Since −1.8 lies in the critical region for the test, we reject the null hypothesis and state that farmers in this region are paying inferior wages to the workers. Notice that if we had made α smaller, such as $\alpha = 0.01$, we could not have rejected the null hypothesis.

Now for a two-tailed test. The regulatory agency that administers the pesticide applicator's test has determined that the average score is $\mu_o = 70$, with standard deviation, $\sigma = 15$. The agency develops a new, multimedia presentation and gives it to a randomly selected group of forty-nine new license applicants to see if the new presentation causes a change in the mean score. The average score for this group is $\overline{X} = 72$. Test at the $\alpha = 0.05$ level of significance.

To solve this example, we first write the null and alternative hypotheses:

$$H_o\text{: } \mu = 70$$
$$H_a\text{: } \mu \neq 70.$$

Next, we divide α in two and obtain the two values of Z from Appendix table 7 that correspond to $\alpha/2 = 0.025$. These are ±1.96. We then compute Z as:

$$Z = \frac{\overline{X} - \mu_o}{\sigma/\sqrt{n}} = \frac{72 - 70}{15/\sqrt{49}} = 0.93$$

Thus, when we compare the computed Z value of 0.93 to the table values of ±1.96, we see that the computed Z is between them and does not fall in the critical region (figure 8.3). So we cannot reject the null hypothesis. In the context of this example, the multimedia presentation does not change the average score on the pesticide applicator's test. The observed sample mean is higher simply because of sampling variation.

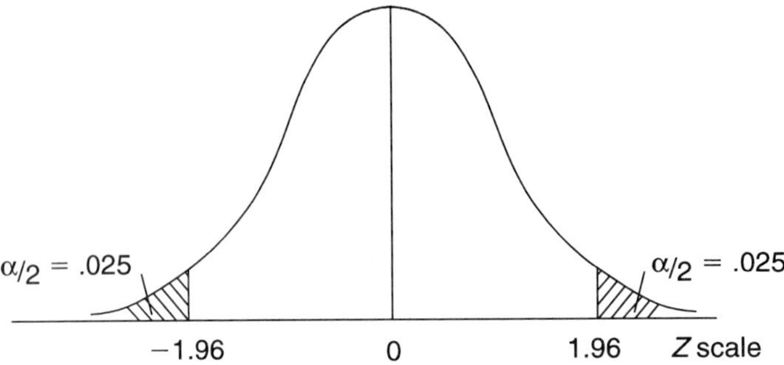

FIGURE 8.3 Sampling distribution and critical region for two-tailed test of μ

One-Sample Tests for μ with σ Unknown

In this section, we again want to test hypotheses about μ, but in this case we do not know the population standard deviation σ. We must estimate it with S, the sample standard deviation. The sampling distribution for $\bar{X}$ in this instance is the t distribution with $(n - 1)$ degrees of freedom, ν.

Large Samples

If we conduct the hypotheses tests with a large sample of data ($n \geq 30$), then we may use the normal distribution as an approximation to the t distribution and conduct the test in the same manner as done in the previous section. There is one change. The computed Z formula contains S in the denominator in the place of σ. Thus, it is not necessary to provide an example for this case.

Small Samples

When we use a small sample ($n < 30$) as the basis for a test of hypotheses on μ, then the t distribution is the sampling distribution for the test. Consider the following example.

A fruit canner maintains on the average 15.33 ounces of peaches in its containers. A random sample of eight cans is selected, and the sample average is computed as 15.15 ounces with $S = 0.20$ ounces. Does this sample indicate at the $\alpha = 0.01$ level of significance that the average weight is being maintained?

The null and alternative hypotheses for this test follows. This is two-tailed test because there could be weight variation in both directions from the standard for the peaches.

$$H_o\colon \mu = 15.33$$
$$H_a\colon \mu \neq 15.33$$

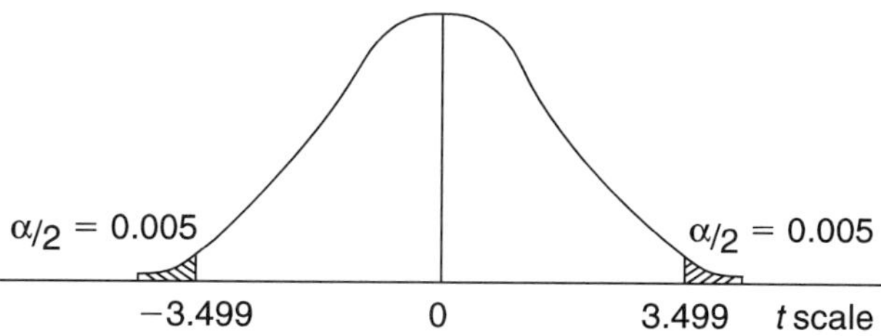

FIGURE 8.4 Critical region for two-tailed t test on μ

For this test, $\alpha = 0.01$, $\nu = 7$, and the t values from Appendix table 8 for a two-sided test are ± 3.499. Thus, the critical region for the test is set up as shown in figure 8.4. We use the formula for computed t to compute the test statistic,

$$t = \frac{\overline{X} - \mu_o}{\frac{S}{\sqrt{n}}} = \frac{15.15 - 15.33}{\frac{0.2}{\sqrt{8}}} = -2.54$$

which when compared to the table values of t, does not fall in the critical region. Thus, we cannot reject the null hypothesis. In the context of this example, the sample mean weight is not different from the standard of 15.33 ounces except for sampling variability, and the process is putting the correct weight of peaches into its containers.

One-Sample Test for p

When we wish to test a hypothesis about the binomial parameter p, we use the normal approximation to the binomial and conduct the test in much the same way as the test of "fairness" for a die, as described earlier in the section on Type I and Type II errors.

Two-Sample Tests for Means with σ_1, σ_2 Known

When sampling is from two populations and we know both population standard deviations, we base tests of hypotheses on the normal probability distribution with random variable $(\overline{X}_1 - \overline{X}_2)$. We set up the test in the same way as a one-sample test except that the hypotheses contain statements about the difference in the population means and we base the computed Z formula on differences, as shown in equation 8.4. The null hypothesis for most two-

$$Z = \frac{(\overline{X}_1 - \overline{X}_2) - (\mu_1 - \mu_2)}{\sqrt{\frac{\sigma^2_1}{n_1} + \frac{\sigma^2_2}{n_2}}} \quad \textbf{8.4}$$

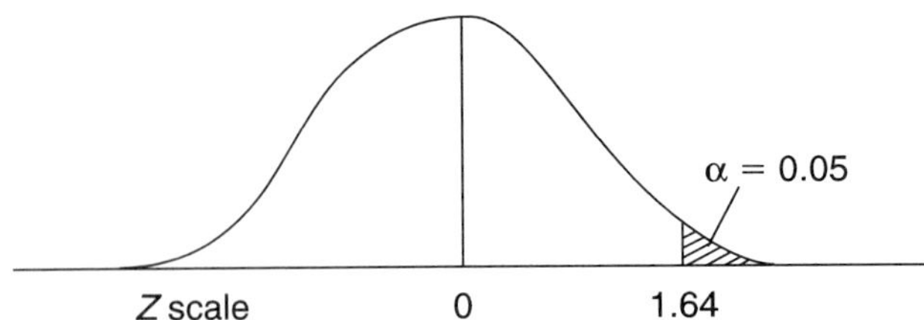

FIGURE 8.5 Critical region for two-sample test of means

sample tests is that there is no difference in the population means, i.e., that $(\mu_1 - \mu_2) = 0$. So zero is generally the value used in the computed Z formula for this difference.

Consider the following example. An experiment station scientist knows that the standard deviation of wheat response to a standard pelletized fertilizer formulated for the region is 0.6 bushels, and the standard deviation for an enhanced formula is 0.5 bushels. The scientist tests a new variety of wheat to determine its response to the two fertilizer formulas by dividing a research plot into fifty squares and treating half of each square with the standard fertilizer, and half with the enhanced formula. At the end of the season, wheat harvested produced an average of 40 bushels with the standard formula and 42 bushels with the enhanced formula. Test at the 0.05 level of significance whether the enhanced formula of fertilizer increases average yield.

To solve this problem, we write the hypotheses as:

$$H_o\colon \mu_1 - \mu_2 = 0$$
$$H_a\colon \mu_1 - \mu_2 > 0$$

We can make this a one-tail test to the right by selecting the sample for the enhanced fertilizer as sample one. In this case, the critical Z value from Appendix table 7 is 1.64 (figure 8.5).

$$Z = \frac{(\overline{X}_1 - \overline{X}_2) - (\mu_1 - \mu_2)}{\sqrt{\frac{\sigma^2_1}{n_1} + \frac{\sigma^2_2}{n_2}}} = \frac{(42 - 40) - 0}{\sqrt{\frac{0.5^2}{25} + \frac{0.6^2}{25}}} = 12.8$$

Computed Z is 12.8, a very large value that falls in the critical region for the test. Thus, we reject the null hypothesis and conclude that wheat yield for this variety is greater with the enhanced formula fertilizer.

Two-Sample Tests for Means with σ_1, σ_2 Unknown

When we do not know the population standard deviations, the t distribution becomes the sampling distribution for the differences in the sample means. However, for large samples, we may employ the normal distribution to approximate the t and use computed Z as the test statistic. We have almost the same formula

TABLE 8.2 Data for Steer Ration Test

Diet	New Protein Supplement	Regular Ration
Average gain ($\bar{X}$), pounds	120	101
$\Sigma(X - \bar{X})^2$	5,032	2,552

as that found in the previous section, except that we substitute S_1^2 in place of σ_1^2 and we replace σ_2^2 with S_2^2, as shown in equation 8.5. Otherwise, we conduct the test in the

$$Z = \frac{(\bar{X}_1 - \bar{X}_2) - (\mu_1 - \mu_2)}{\sqrt{\frac{S^2_1}{n_1} + \frac{S^2_2}{n_2}}} \quad \boxed{8.5}$$

same manner as in the previous section.

Small, Independent Samples

When both samples are small ($n_1, n_2 < 30$) and the populations are independent, we base the hypotheses tests on the t distribution with $(n_1 + n_2 - 2)$ degrees of freedom and calculate computed t as in equation 8.6. Consider the following example. Two pens of steers are fed two

$$t = \frac{(\bar{X}_1 - \bar{X}_2) - (\mu_1 - \mu_2)}{\sqrt{\frac{S^2}{n_1} + \frac{S^2}{n_2}}} \quad \textit{where } S^2 = \frac{(n_1 - 1)S_1^2 + (n_2 - 1)S_2^2}{n_1 + n_2 - 2} \quad \boxed{8.6}$$

different rations for a given period of time. The twelve steers in the first pen are given a ration with a new protein supplement, while the seven steers in the second pen are given a standard ration (table 8.2). Is the new protein supplement more effective than the standard ration at the 5 percent level of significance?

To perform this test, we set up the null and alternative hypotheses as:

$$H_o\text{: } \mu_1 - \mu_2 = 0$$
$$H_a\text{: } \mu_1 - \mu_2 > 0$$

and we compute degrees of freedom as $\nu = (n_1 + n_2 - 2) = (12 + 7 - 2) = 17$. Thus, since this is a one-sided test to the right, we find table t for $\alpha = 0.05$ in Appendix table 8 is 1.74. We can use the fact that $S^2 = \Sigma(X - \bar{X})^2/(n - 1)$ to understand that $\Sigma(X - \bar{X})^2 = (n - 1)\ S^2$. Thus, we may add these two "sums of squares" and divide their sum by degrees of freedom to obtain an estimate of the pooled variance that we need for the computed t formula. Thus,

$$S^2 = (5{,}032 + 2{,}552)/17 = 446.12.$$

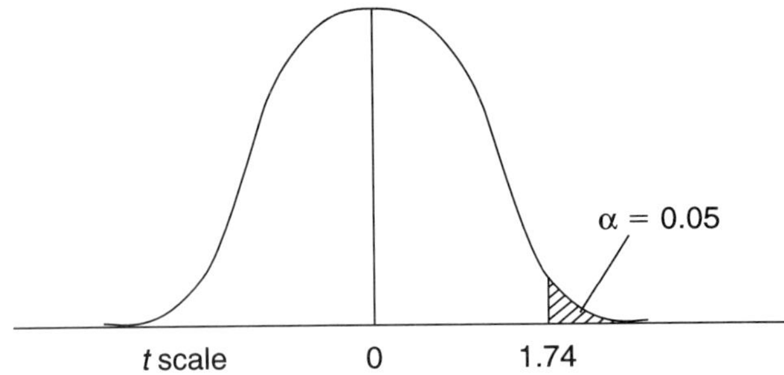

FIGURE 8.6 Critical region for steer ration test

When we substitute this and other values in the computed t formula, we get:

$$t = \frac{(120 - 101) - 0}{\sqrt{\frac{446.12}{12} + \frac{446.12}{7}}} = 1.89$$

Since the computed t falls in the critical region for the test, we reject the null hypothesis. In the context of this problem, the ration with the new protein supplement increases average weight gain for the steers over the standard ration (figure 8.6).

Small, Paired Samples

When the samples are small and not independent, then we use the paired t to test hypotheses between two means. Degrees of freedom is the number of pairs less one ($n - 1$) and the sample statistic is the mean of the differences between the samples, as indicated in equation 8.7

$$\bar{d} = \Sigma d/n. \quad \boxed{8.7}$$

Consider the following example. Litter mates from five different litters of pigs were fed two different rations. The data are given in table 8.3 as weight gain

TABLE 8.3 Weight Gain in Pounds for Litter Mates from Five Pig Litters Fed Two Rations

Pair	Ration 1	Ration 2
1	25	19
2	30	32
3	28	21
4	34	34
5	23	19

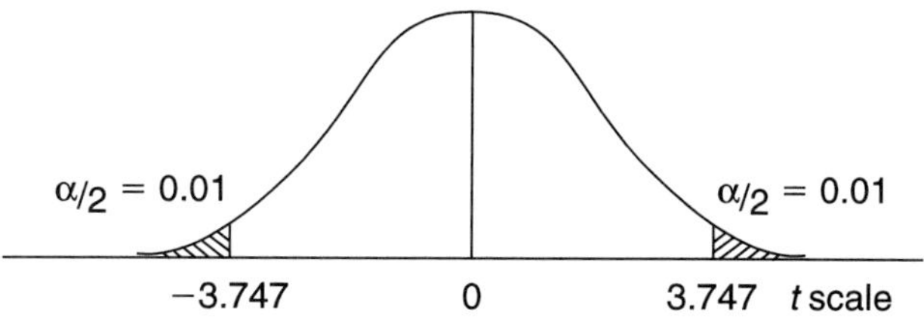

FIGURE 8.7 Critical region for paired *t* test

in pounds. At the 2 percent level of significance, can either ration be considered superior?

To perform this test, we first write the null and alternative hypotheses as:

$$H_o\text{: } \mu_1 - \mu_2 = 0$$
$$H_a\text{: } \mu_1 - \mu_2 \neq 0$$

The degrees of freedom $\nu = (5 - 1) = 4$, and the t value for $\alpha/2 = 0.01$ from Appendix table 8 is ± 3.747 (figure 8.7). We obtain the computed t after we calculate $\bar{d}$ and S_d. The differences for the pairs are $\{6, -2, 7, 0, 4\}$, which gives rise to a $\bar{d} = 3$, and $S_d = 3.87$. Thus, computed t is

$$t = \frac{\bar{d} - (\mu_1 - \mu_2)}{\frac{S_d}{\sqrt{n}}} = \frac{3 - 0}{\frac{3.87}{\sqrt{5}}} = 1.73$$

Because the computed t does not fall in the critical region for the test, we cannot reject the null hypothesis. Therefore, we find no difference in the rations with respect to weight gain for the pigs in this test.

Two-Sample Tests for Proportions

When we have two samples and wish to test the difference between population proportions, we may use the Z test, which is the normal approximation to the binomial distribution. We set up the null and alternative hypotheses in the same way as we did for two means, i.e., we subtract the two proportions and state that their difference is zero under the null and not zero under the alternative. The computed Z is the difference in the two sample proportions minus the difference in the population values under the null hypothesis (zero) divided by the standard error of the difference in the proportions. In computing the standard error, we use all of the sample information, since if the null hypothesis is true, the two samples represent information from the same population. Thus, we compute our estimate of the population proportion by equation 8.8

$$\hat{p} = \frac{n_1\hat{p}_1 + n_2\hat{p}_2}{n_1 + n_2} \tag{8.8}$$

and use it to compute the standard error of the difference in equation 8.9. The computed Z for the

$$S_{\hat{p}_1 - \hat{p}_2} = \sqrt{\frac{\hat{p}\hat{q}}{n_1} + \frac{\hat{p}\hat{q}}{n_2}} \tag{8.9}$$

test is given by equation 8.10. We compare computed Z to the critical value of Z from Appendix table 7.

$$Z = \frac{(\hat{p}_1 - \hat{p}_2) - (p_1 - p_2)}{S_{\hat{p}_1 - \hat{p}_2}} \tag{8.10}$$

Consider the following example. You are a marketing researcher for a food company and want to see if your ad campaign has changed the proportion of consumers who prefer your orange juice. You survey 400 people in the city where the ad campaign is running and find that 192 preferred your orange juice. Also, you survey 500 people in another city that has not had the campaign and 210 preferred the orange juice. Test at the 5 percent level of significance to see if the ad campaign is effective. We set up the null and alternative hypotheses as:

$$H_o: p_1 - p_2 = 0$$
$$H_a: p_1 - p_2 > 0$$

Since this is a one-sided test, our critical $Z = 1.64$ from Appendix table 7. We compute our estimate of the population proportion using equation 8.8 as $\hat{p} = (192 + 210)/(400 + 500) = 0.447$. Thus, we can calculate the standard error of the difference in the proportions using equation 8.7 as $\sqrt{(0.447)(0.553)/400 + (0.447)(0.553)/500} = 0.0333$. Thus, our computed Z becomes $[(0.48 - 0.42) - (0)]/0.0333 = 1.80$. Since computed Z exceeds the table value of 1.64, we reject the null hypothesis and state that the proportion of people who prefer our orange juice is higher in the city where the ad is running. Thus, the ad appears to be effective.

Determining Sample Size in Surveys

We have already discussed the advantages of taking samples and using them as the basis for statistical inference. However, in many situations we do not know the best size of sample to select. In such cases, we specify the amount of error we are willing to accept and calculate the sample size based on it.

For Statistical Inference on μ

When we are concerned with making an inference about the population mean, μ, the amount of error that we have is given by the distance $(\bar{X} - \mu)$, which is the amount by which the estimate, $\bar{X}$, misses the population parameter, μ, due to sampling variation. If we set this error to a specific value, say E, then we can manipulate the Z formula to solve for the sample size necessary to provide this error, provided we also set the confidence level for the inference, say at 99 percent, 95 percent, etc. We recall that Z is computed as in equation 8.11 and if the required confidence

$$Z = \frac{\overline{X} - \mu}{\frac{\sigma}{\sqrt{n}}} \tag{8.11}$$

level is $(1 - \alpha)$, then we can write the confidence interval as in equation 8.12. Since the normal

$$-Z_{\alpha/2} \cdot \sigma/\sqrt{n} \leq \overline{X} - \mu \leq Z_{\alpha/2} \cdot \sigma/\sqrt{n} \tag{8.12}$$

distribution is symmetric, we can concentrate on the inequality to the right, which gives the expression in equation 8.13.

$$\overline{X} - \mu \leq Z_{\alpha/2} \cdot \sigma/\sqrt{n} \tag{8.13}$$

It says that the largest value that $\overline{X} - \mu$ can assume is $Z_{\alpha/2} \cdot \sigma/\sqrt{n}$, but we have already called the largest sampling error that we want $E = \overline{X} - \mu$. Hence, we can put these together and write the equation for the sampling error (equation 8.14). By solving equation 8.14

$$E = Z_{\alpha/2} \cdot \sigma/\sqrt{n} \tag{8.14}$$

for n, we get the sample size, which assures us, with 100 $(1 - \alpha)$ percent confidence, that $\overline{X} - \mu$ will be no larger than the sampling error E. Thus, our equation for the optimal sample size is given in equation 8.15.

$$n = \frac{Z^2_{\alpha/2}\sigma^2}{E^2} \tag{8.15}$$

Consider the following example. The Greenhouse Company erects plastic frame Quonset-style houses covered with a translucent plastic. It is considering purchasing its frame members from a new plastics company that has located near its plant. It knows from past experience that the standard deviation of the length of frame members is 0.2 feet. How large a sample is needed to estimate the population mean length of the frame members within 0.03 feet with a 95 percent confidence estimate?

For this problem, $\sigma = 0.2$, $E = 0.03$, and $Z_{\alpha/2} = 1.96$. Substituting these values into the equation for n gives:

$$n = \frac{Z^2_{\alpha/2}\sigma^2}{E^2} = \frac{1.96^2\,0.2^2}{0.03^2} = 170.7$$

Thus, the Greenhouse Company should take a sample of 171 frame members from the new supplier to estimate their average length within 0.03 feet with 95 percent confidence.

For Statistical Inference on p

We can determine the size of sample for the binomial parameter p if the maximum sample error we are willing to accept, $E = \hat{p} - p$, is specified in advance along with the confidence level for the test. In many instances, there is no value of $\hat{p}$ available to use in calculating σ_p since the sample has not yet been taken. In these cases, we use an estimate of $p = ½$ since it can be proved that the value of σ_p will be largest when $p = ½$. So if we substitute $p = ½$ into the Z equation, in which $Z = (\hat{p} - p)/\sqrt{pq/n}$ to get $Z = (\hat{p} - p)/\sqrt{(1/2)(1/2)/n}$ and solve for n, we obtain the formula we are seeking in equation 8.16.

$$n = (p)(1 - p)\frac{Z^2{}_{\alpha/2}}{E^2} = \frac{Z^2{}_{\alpha/2}}{4E^2} \quad \boxed{8.16}$$

Consider the following example. Suppose the Melon Company, a watermelon packer, wishes to obtain an estimate of the proportion of cull melons that go through the plant that is accurate within 3 percentage points with 95 percent confidence. How large a sample must be taken?

We have $E = 0.03$, $Z_{\alpha/2} = 1.96$. Thus, $n = 1.96^2/(4)(0.03)^2 = 1{,}067.1$, or, 1,068 melons. The sample size could be reduced somewhat if we had an estimate of p, the proportion of cull melons from a previous time period, etc.

Sample Size to Control Both α and β

In testing hypotheses, we often want to set α at some predetermined level such as $\alpha = 0.05$, and we also wish to set β for a given value of H_a. We can do this provided we compute the sample size that controls both errors at their desired level and use that size sample when we do the test.

Recall that when we set up a test of hypotheses, we select a decision rule based on a critical value of Z that cuts off the probability α in one tail of the sampling distribution for the null hypothesis. Once we establish the decision rule, then we calculate β as the portion of the sampling distribution under the alternative hypothesis that crosses the decision line and lies in the acceptance region for the null hypothesis. Thus, we can formulate the critical value for the decision rule with two expressions (equations 8.17 and 8.18)—one based on the sampling distribution when H_o is true, and another based on the sampling distribution when H_a is true. Both of these statements contain n, which we can solve for provided we set β at some predetermined

$$\bar{X}_C = \mu_o - Z_\alpha \cdot \sigma/\sqrt{n} \quad \boxed{8.17}$$

$$\bar{X}_C = \mu_a + Z_\beta \cdot \sigma/\sqrt{n} \quad \boxed{8.18}$$

level. These two statements indicate a one-sided test to the left. For a one-sided test to the right, we reverse the plus and minus signs in the two expressions, i.e., the first equation contains the plus sign; the second the minus sign. For a two-sided test, we substitute $Z_{\alpha/2}$ for Z_α in equation 8.17. Since the two expressions are both equal to $\bar{X}_C$, we can set them equal to each other and solve for n to obtain what we want (equation 8.19). We use positive Z values in equation 8.19

$$n = \left[\frac{(Z_\alpha + Z_\beta)\sigma}{(\mu_o - \mu_a)}\right]^2 \quad \boxed{8.19}$$

since the signs of the equations for $\overline{X}_C$ cause us to get the correct values.

Consider the following example. A large agribusiness company wishes to survey the market for new college graduates with a B.S. in Agribusiness Management. It believes that the market has not changed and desires to test the null hypothesis that the average beginning salary is $2,250 per month with $\alpha = 0.05$. It also wants to be able to detect a difference of $100 per month with probability $\beta = 0.03$. If the standard deviation of the population of beginning salaries is $\sigma = \$300$, how large a sample must it take? This is a two-tailed test with the hypotheses:

$$H_o: \mu = \$2{,}250$$
$$H_a: \mu \neq \$2{,}250$$

Thus, we divide α in half and look up $Z_{\alpha/2}$ in Appendix table 7 as 1.96. Since $\beta = 0.03$, then $Z_\beta = 1.88$ from the same table. We want to know if starting salaries have changed by $100 per month, so the relevant value of the alternative hypothesis is different by $100, i.e., $(\mu_o - \mu_a) = 100$. We now have the values for our formula needed to solve for n.

$$n = \left[\frac{(Z_\alpha + Z_\beta)\sigma}{(\mu_o - \mu_a)}\right]^2 = \left[\frac{(1.96 + 1.88)(300)}{(100)}\right]^2 = 132.7$$

Thus, the company needs a sample size of 133 newly employed Agribusiness Management graduates.

Appendix to Chapter 8 Using Excel to Perform Two-Sample Tests of Hypotheses

Two-Sample *Z* Test

To perform the preceding test using a computer program such as Excel, place the first mean in cell A1, the second in cell B1, and click on the Data Analysis submenu of the Tools Menu and select the z-Test: 2 Sample for Means procedure. In the table that pops up, write the range, A1, for the first mean in the top box, and the range, B1, for the second mean in the second box. Place a zero in the mean difference box, and calculate the known variances as σ^2/n and place them in their respective boxes. Observe the value of alpha and change it if necessary. Finally select an output range such as D1 and click OK. The output will appear on the spreadsheet as in table 8.4. To read the first column, we must increase its width to 20 spaces or so. Notice that the procedure assumes that we

TABLE 8.4 Spreadsheet output for two-sample Z test

z-Test: Two-Sample for Means

	Variable 1	Variable 2
Mean	42	40
Known variance	0.01	0.0144
Observations	1	1
Hypothesized mean	0	
Difference		
z	12.80369	
$P(Z <= z)$ one-tail	0	
z Critical one-tail	1.644853	
$P(Z <= z)$ two-tail	0	
z Critical two-tail	1.959961	

have the raw data in columns A and B of the spreadsheet. Since we do not, we must calculate the known variances in the manner we used. If there were raw data in those columns, we could just place the variance values there as opposed to the values of σ^2/n.

Independent t Test

This procedure requires that we have the raw data as opposed to summary data of the type we used in the preceding example. Therefore, consider the following example. Twenty experimental plots were planted to corn. On half of the plots, a new fertilizer formula was applied at the standard rate; on the other half, the recommended formula was used at the same rate. Yields per acre are given in the following table. Is there evidence that the new formula is better at the 1 percent level of significance?

New formula	85	91	86	102	94	98	88	105	112	82
Old formula	82	95	80	97	91	90	87	100	103	80

To do the analysis with Excel, input the data on the spreadsheet in columns A and B with the variable titles in cells A1 and B1 and then select the Data Analysis submenu of the Tools Menu and choose the t-Test: Two-Sample Assuming Equal Variances procedure. On the pop-up table, input the data ranges for the first and second variables (A1:A11 for variable 1 and B1:B11 for variable 2), put 0 in the box for mean difference, check the labels box, change the value of alpha to 0.01, and select the output range as the cells beginning with D1 then click OK.

In examining the output (table 8.5), we see that the computed t is 0.9394, which is significant at only the 0.17 level of significance, not 0.01 as we specified,

TABLE 8.5 Spreadsheet output for two-sample *t* test

***t*-Test: Two-Sample Assuming Equal Variances**		
	New Formula	**Old Formula**
Mean	94.3	90.5
Variance	95.34444444	68.27777778
Observations	10	10
Pooled variance	81.81111111	
Hypothesized mean	0	
Difference		
df	18	
t Stat	0.939425735	
$P(T <= t)$ one-tail	0.179974319	
t Critical one-tail	2.552378646	
$P(T <= t)$ two-tail	0.359948638	
t Critical two-tail	2.878441592	

so we cannot reject the null hypothesis. The critical value of t is 2.55, which is larger than our computed value. There are 18 degrees of freedom for this test. The mean corn yield for the new formula is 94.3 bushels/acre, while the average for the old formula is 90.5. The pooled variance is 81.81.

Exercises

1. What is the difference between a parameter and a statistic?
2. Analyze the following statement: In statistical hypothesis testing, we should always set the level of significance (or α) very low, say at 0.01, so we won't make many errors.
3. State the null and alternative hypotheses in each of the following:
   a. A cereal company wants to test its production process against the claim by a leading consumer organization that it puts less than 14 ounces of cereal in its 14-ounce boxes of Wakem-Up Cereal.
   b. The Farm Board claims that the percentage of farmers who favor stricter environmental controls is the same in Oklahoma and New Jersey.
   c. A farm shop teacher doubts the manufacturer's claim that the average width of Black Mule welding gloves is 2 inches.
   d. An agricultural economist wishes to test whether the percent of unemployed people in rural areas is less than it was last year, given that the total economy is growing.
   e. A farm supply store manager wants to check the assumption that the average credit account balance at his business is at least $70.
4. Criticize the following:
   a. Because it is more important to control Type I errors than Type II, we should design our tests to be sensitive only to Type I errors.
   b. Since a decrease in a Type I error increases a Type II, it is impossible to control both.
   c. Once we have made a decision in a test of hypotheses, it is possible that both Type I and Type II errors have occurred.
5. The specifications for a component for a cotton harvester state that the mean length should not be less than 100 inches with a standard deviation of 8 inches. You select a random sample of eighty-one components from a very large shipment and compute the sample mean and standard deviation as 98.5 and 8 inches, respectively.
   a. Based on your analysis, would you accept this shipment? Justify your answer.
   b. Given your critical value in part a., what is the probability of a Type II error for a shipment of components having a mean length of 97 inches and a standard deviation of 8 inches?
6. A standardized farm skills test has been given for several years to high school agriculture students. It has a mean of 75 and a standard deviation of 7. A teacher decides to try to raise student scores by increasing emphasis on welding. Thus, the teacher's next class of twenty-five students receives instruction with increased welding emphasis and scores 78 on the test. Can we say at the 5 percent level of significance that the instruction with welding emphasis increased average test scores?
7. When injected into chickens, a certain growth hormone is said to increase average weight of broilers by 0.5 pound. A quality control expert for

Howdy Farms obtains a random sample of nine broilers injected with the hormone and calculates the average weight gain to be 0.6 pound with $S = 0.2$. If we do a two-tailed test with $\alpha = 0.05$, will it confirm that the average weight increase is 0.5 pound?

8. Two horticulturists score plants for viability. The first scores twelve plants with an average of 90 while the second scores eight plants with an average of 85. Past experience with these two professionals indicates that each scores plants with a variance of 40. Test at the 5 percent level to see if there is a difference in their average scores.

9. We want to test the acceptability of a new fruit product. If half of the consumers tested say they will buy it, we will market it; otherwise we will not. We select a random sample of sixty-four consumers and twenty-six state they will buy it. Test at the 5 percent level of significance to see if we should market the new fruit product.

10. Two new liquid media are being tested for growing tomatoes hydroponically. The first media is applied to ten plants that produce an average of twenty-eight tomatoes with a standard deviation of 2, while the second is applied to fifteen plants of the same variety that produce an average of eighteen tomatoes with a standard deviation of 9. Test whether the first media is significantly better than the second at the 5 percent level of significance.

11. A rancher has two places separated by several miles and two cattle herds. On the home place, the herd is managed so that there will be a fall calf crop, and on the other place, a spring calf crop. The home place herd consists of 420 cows that produced 378 calves last fall, while on the other place, there are 110 cows that produced 90 calves in the spring. Test at the 5 percent level of significance to see if the calving percentage is the same at both places.

12. Eight agriculture students are given a pretest and then a post-test after receiving instruction on grafting and layering. Their scores follow. Test at the 1 percent level of significance to see if the instruction was effective.

Student	Pretest Score	Post-Test Score
1	37	75
2	62	87
3	71	95
4	45	72
5	53	78
6	24	63
7	90	98
8	86	84

13. An agricultural economist wants to determine the average wage rate for tobacco laborers in the flue-cured growing areas of the Carolinas, Georgia, and Florida. She does a pilot study that provides a mean of \$6.80 and a standard deviation of \$0.50. If she wants to be 95.5 percent confident that the maximum error in her wage estimate is no more than $.10, how large a sample should she take to do the complete study?

14. A large shipment of grain is to be received at a terminal elevator and is to be sampled to determine the percent trash content. How large a random sample should be drawn if we want to estimate the percent trash to within 1.5 percentage points with 99 percent confidence? Assume the proportion of trash content is no larger than 0.10 based on past experience with this grain.
15. Solve exercise 12 with Excel.
16. Solve the following with Excel. A farmer wants to insulate his shop. In deciding which material to use, he comes across a study that has provided data on the impact strength of two materials as follows. But there has been no analysis. Test at the 5 percent level of significance to see which material has the higher impact strength.

Insulation 1	1.50	1.41	1.58	1.40	1.45	1.57	1.53	1.46
Insulation 2	1.14	1.22	1.20	1.19	1.27	1.24	1.31	1.23

CHAPTER

Analysis of Variance

In the previous chapters on statistical inference, we presented tests of hypotheses about values of parameters from one and two populations. In this chapter, we consider procedures for testing hypotheses about parameters from three or more populations. This topic is necessary because pairwise comparisons are cumbersome and do not always lead to correct results, i.e., if we have a test of three means, we can test all possible pairs of means using procedures already outlined, but the tests may not be reliable. The analysis of variance offers an excellent procedure for making tests about three or more means that is reliable.

The analysis of variance has its own terminology that is somewhat different, so we will first explore terms. We use sample data to compare several levels of a random variable called a treatment. A **treatment** refers to any random variable that the researcher controls. For example, an entomologist might test the effectiveness of several levels of an insecticide in controlling a certain moth on apples as a treatment. Or an animal scientist might select various concentrations of protein in a beef-feeding trial as a treatment. A second term is experimental unit. An **experimental unit** is the entity that receives a treatment. For the entomologist, the experimental unit may be a tree or a plot of land of a certain size in an apple orchard. For the animal scientist, the experimental unit may be a 600-pound steer or a bull. A statistical term that may be new is **mean square,** which is another name for variance. In presenting the results of analysis of variance, we generally construct a table with a column titled "mean square," which contains the variances for the various components of the analysis.

The focus of the analysis of variance is upon variances. We recall that we can write a variance as the appropriate sum of squares divided by degrees of freedom. In analysis of variance, the total sample variance is partitioned. If the null hypothesis is true, the partitions provide independent estimates of the same variance, and we compute an F statistic as a ratio of the two estimates. The sampling distribution of the ratio of variances is continuous and forms the

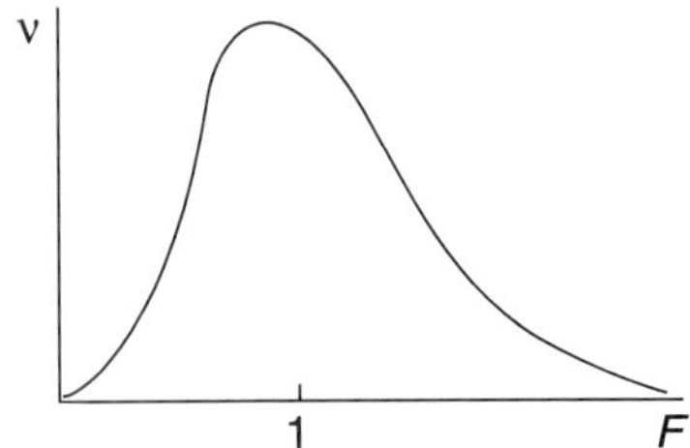

FIGURE 9.1 The *F* distribution

F distribution (figure 9.1), named for its developer, Sir R. A. Fisher. The random variable *F* is always in the range of zero to infinity since a ratio of squared numbers cannot be negative. It is skewed positively, but approaches symmetry as the degrees of freedom for each variance approach infinity. Thus, we expect it to be highly skewed for small samples and less skewed as sample sizes increase. Degrees of freedom determine its shape. See Appendix table 10 for a listing of critical values of *F* for only two values of α—0.05 and 0.01. The values in the table in regular print refer to table *F* when $\alpha = 0.05$, while the values in bold print refer to table *F* when $\alpha = 0.01$. The column headings along the top are degrees of freedom for the variance in the numerator of the *F* ratio, while the row headings along the side refer to degrees of freedom for the variance in the denominator.

The theoretical concepts of analysis of variance are based on a set of assumptions:

1. The *k* samples are drawn randomly from *k* populations.
2. The *k* populations are normal.
3. The variances of the *k* populations are equal.

Also, the *k* population means must be equal for *F* to be a random variable in the *F* distribution. But this condition is the hypothesis we are testing by the analysis of variance. We consider the hypothesis as true before the test, but may reject it after the test. The other three conditions must be true both before and after the test. To assure that the first assumption is true, we must be satisfied that the samples are randomly selected using one of the procedures we discussed earlier for choosing random samples. This is the most important assumption, and violation of the randomness assumption causes serious error in the *F* test. The second assumption is not so critical. Non-normality of the populations causes the true critical values of *F* to be larger than those in the appendix table. The third assumption about equality or homogeneity of variances of the populations being sampled is also not critical. When variances are not too much different, then the *F* test is not seriously affected, and this effect can be reduced by making the sample sizes equal. Statistical procedures are available to us for dealing with departures from normality and for heterogeneity of variances.

The analysis of variance is a very comprehensive topic in statistics. In the sections that follow, we discuss analysis of variance techniques for only three experimental designs.

One-Way Analysis of Variance

In a one-way analysis of variance, we apply several levels of a treatment and test the null hypothesis that the treatment means are equal against the alternative that they are not all equal. This is equivalent to saying in the null hypothesis that the variance of the treatment means is zero against the alternative hypothesis that the variance of the treatment means is greater than zero. Thus, we have:

$$H_o\text{: } \mu_1 = \mu_2 = \dots = \mu_k \text{ (or } \sigma_\mu^{\ 2} = 0)$$
$$H_a\text{: } \mu_1 \neq \mu_2 \neq \dots \neq \mu_k \text{ (or } \sigma_\mu^{\ 2} > 0)$$

and we reject the null hypothesis whenever the computed F exceeds the table value for α. When the means are equal, the k samples can be treated as one sample from one single population, but when the means are not equal, then they must be recognized as k independent random samples from k different populations. We stated under the assumptions that we first conduct the analysis of variance as if the null hypothesis is true. Thus, we begin by dealing with the data as if it comes from one large, homogeneous population. We take the total variation and partition it into two components—one for treatments and another for experimental error. If the null hypothesis is true and the treatment means are equal, then the component of the total variation for treatments will contain no variation due to differences in the treatment means; rather it will be just another estimate of experimental error. If we make a ratio of the two variances—treatment and experimental error under these conditions—we should obtain a number for computed F close to one since they are estimates of the same quantity. If, however, the treatment means are not equal, then the component of the total variation for treatments will contain both variation due to differences in the treatment means and due to experimental error, and it will inflate. In this case, a ratio of the two variances, treatment and experimental error, will generate a large number for computed F—greater than the table value—and will lead to the rejection of the null hypothesis.

We perform this test for two cases: equal size samples and unequal sample sizes.

Equal Sample Size

As an illustration, consider the data in table 9.1 on methods of shipping eggs under evaluation by an egg wholesaler. Four methods are tried, with five lots of eggs shipped by each method. The number of broken eggs in each lot is recorded in the table. Is the average number of broken eggs the same for each shipping method? Test at the 1 percent level of significance.

TABLE 9.1 Sample Number of Broken Eggs for Four Shipping Methods

Sample Lot Number	Shipping Method *A*	*B*	*C*	*D*
1	4	9	1	9
2	3	7	6	7
3	1	6	4	7
4	2	8	3	6
5	5	8	4	8
Totals	15	38	18	37
Means	$\bar{X}_{.1} = 3.0$	$\bar{X}_{.2} = 7.6$	$\bar{X}_{.3} = 3.6$	$\bar{X}_{.4} = 7.4$

$$\bar{X}_{..} = \frac{15 + 38 + 18 + 37}{5(4)} = 5.4$$

The analysis of variance procedure used to determine whether shipping method affects number of eggs broken involves two variables. The shipping method is a qualitative variable and has four values, one for each treatment. The number of broken eggs is a response variable, which is always quantitative since the response achieved may depend upon the particular treatment used. Although analysis of variance is primarily concerned with qualitative factors, we may also use it with quantitative variables if we fix them at a few key levels to create quantitative categories rather than continuous variables.

To discuss the data in table 9.1, we must first establish some notation. We represent each sample number of broken eggs by X_{ij}, where i refers to the row or observation number in the sample and j refers to the column. For example, X_{21} = three broken eggs—the number broken in the second lot sent by shipping method *A*, the first treatment. We can calculate the sample mean for the jth treatment column as $\bar{X}_{.j}$ (*x* bar dot j) $= \Sigma X_{ij}/n$, where we sum over i as we indicate by the dot in the place of i in the subscript for $\bar{X}$. In our example, $j = 1, 2, 3$, or 4, depending upon which treatment we are considering. For the first treatment, we compute the sample mean by summing the values in column A and dividing by $n = 5$:

$$\bar{X}_{.1} = \frac{4 + 3 + 1 + 2 + 5}{5} = 3.0 \text{ broken eggs}$$

We calculate the remaining sample means in the same way: $\bar{X}_{.2} = 7.6$, $\bar{X}_{.3} = 3.6$, and $\bar{X}_{.4} = 7.4$.

The grand mean, $\bar{X}_{..}$ (x bar dot dot) is the average for all twenty observations since if the null hypothesis is true, the entire data set can be pooled. Thus we compute $\bar{X}_{..}$ with equation 9.1

$$\bar{X}_{..} = \Sigma\Sigma X_{ij}/nk \qquad \boxed{9.1}$$

where we sum over both i and j and require two summation signs. We sum over the rows first to obtain column totals and then add these to get the grand total. We divide by the number of rows, n, multiplied by the number of columns, k, so that we use all of the sample observations in the divisor. We compute the grand mean in table 9.1 as $\overline{X}_{..} = 5.4$ broken eggs.

Since we are dealing with several populations, we use the sample data to measure three sources of variability: treatment variation or treatment **sum of squares**, which measures how sample results differ among the various treatments; error variability or error sum of squares, which measures collectively how the observations vary within their respective samples; and total variability or sum of squares, which measures total variation of the sample observations for the whole experiment without regard to the number of populations.

Treatment Sum of Squares

To summarize the variability among the various sample results, we use the treatment sum of squares, which we define as SST with equation 9.2. In calculating SST, we sum squared

$$SST = n\Sigma(\overline{X}_{.j} - \overline{X}_{..})^2 \qquad \boxed{9.2}$$

deviations of the treatment means from the grand mean and multiply by the number of observations per sample, n. Thus for the data in table 9.1, we have:

$$SST = 5[(3.0 - 5.4)^2 + (7.6 - 5.4)^2 + (3.6 - 5.4)^2 + (7.4 - 5.4)^2]$$
$$SST = 5(5.76 + 4.84 + 3.24 + 4.00) = 89.2$$

We multiply the squared deviations from the mean by $n = 5$, the number of observations per sample treatment, so that with $k = 4$ treatments, all of the $nk = 20$ observations are represented. Since SST is obtained as the differences in sample means, it is often called **explained variation.** It explains the differences in the sample means that might be due to actual differences in the treatment populations rather than chance.

Error Sum of Squares

We compute the **variation within samples,** or error sum of squares, by adding the squared deviations of the individual sample observations from their own sample means. Thus, the error sum of squares, SSE, is defined in equation 9.3. We compute this quantity for the

$$SSE = \Sigma\Sigma(X_{ij} - \overline{X}_{.j})^2 \qquad \boxed{9.3}$$

egg-shipping method example in table 9.2 as $SSE = 33.6$. In making this calculation, we follow the same arrangement we used earlier—first we sum the squared deviations for each treatment, then add them to obtain SSE. We also account for each sample observation in the error sum of squares as we did in the treatment sum of squares.

TABLE 9.2 Calculation of the Error Sum of Squares for the Egg-Shipping Method Example

i Value	$(X_{i1} - \bar{X}_{.1})^2$	$(X_{i2} - \bar{X}_{.2})^2$	$(X_{i3} - \bar{X}_{.3})^2$	$(X_{i4} - \bar{X}_{.4})^2$
1	$(4 - 3)^2 = 1.00$	$(9 - 7.6)^2 = 1.96$	$(1 - 3.6)^2 = 6.76$	$(9 - 7.4)^2 = 2.56$
2	$(3 - 3)^2 = 0.00$	$(7 - 7.6)^2 = 0.36$	$(6 - 3.6)^2 = 5.76$	$(7 - 7.4)^2 = 0.16$
3	$(1 - 3)^2 = 4.00$	$(6 - 7.6)^2 = 2.56$	$(4 - 3.6)^2 = 0.16$	$(7 - 7.4)^2 = 0.16$
4	$(2 - 3)^2 = 1.00$	$(8 - 7.6)^2 = 0.16$	$(3 - 3.6)^2 = 0.36$	$(6 - 7.4)^2 = 1.96$
5	$(5 - 3)^2 = 4.00$	$(8 - 7.6)^2 = 0.16$	$(4 - 3.6)^2 = 0.16$	$(8 - 7.4)^2 = 0.36$
Totals	10.00	5.20	13.20	5.20

$SSE = \Sigma\Sigma(X_{ij} - \bar{X}_{.j})^2 = 10.00 + 5.20 + 13.20 + 5.20 = 33.6$

Total Sum of Squares

We obtain the total sum of squares as the sum of squares for a single combined sample, ignoring the treatment groupings (equation 9.4). For the egg-shipping method example,

$$\text{Total } SS = \Sigma\Sigma(X_{ij} - \bar{X}_{..})^2 \qquad \textbf{9.4}$$

we calculate the total sum of squares as

$$\text{Total } SS = (4 - 5.4)^2 + (3 - 5.4)^2 + \ldots + (8 - 5.4)^2 = 122.8.$$

Note that if we add the treatment and error sums of squares, they equal the total sum of squares, (equation 9.5)

$$\text{Total } SS = SST + SSE \qquad \textbf{9.5}$$

and for the egg-shipping method example, we obtain the following results:

$$\text{Total } SS = 89.2 + 33.6 = 122.8.$$

The F *Test*

Once we have computed the sum of squares, we can perform the F test to decide whether to reject the null hypothesis, H_o: $\mu_1 = \mu_2 = \ldots = \mu_k$. The F test is the ratio expressed in equation 9.6,

$$F = \frac{\dfrac{SST}{k - 1}}{\dfrac{SSE}{n - k}} \qquad \textbf{9.6}$$

which follows the F distribution with $k - 1$ degrees of freedom in the numerator and $n - k$ in the denominator, in which k is the number of treatments and n is the number of observations per sample. We can rewrite this expression for F,

TABLE 9.3 ANOVA Table for One-Way Analysis of Variance

Source of Variation	Degrees of Freedom	Sum of Squares	Mean Square	F
Treatments	$k - 1$	$SST = n\Sigma(\bar{X}_{.j} - \bar{X}_{..})^2$	$MST = SST/(k - 1)$	MST/MSE
Error	$k(n - 1)$	$SSE = \Sigma\Sigma(X_{ij} - \bar{X}_{.j})^2$	$MSE = SSE/(k(n - 1))$	
Total	$kn - 1$	$SS = \Sigma\Sigma(X_{ij} - \bar{X}_{..})^2$		

which is a ratio of sums of squares to degrees of freedom, as a ratio of mean squares since the definition of a mean square is sum of squares divided by degrees of freedom. Thus, we have equation 9.7.

$$F = \frac{MST}{MSE} \tag{9.7}$$

The size of the F ratio calculated from the data serves as the basis for rejecting the null hypothesis. If it exceeds the size of the critical value of F_α found in Appendix table 10, then we reject the null hypothesis; otherwise not. We generally present the one-way analysis of variance in an ANOVA table, such as that shown in table 9.3. Thus, for the egg-shipping method example, the ANOVA is given in table 9.4. We compare computed $F = 29.73/2.1 = 14.16$ from table 9.4, with F_α or critical $F = F_{0.01} = 5.2922$. Since computed F from table 9.4 is larger than the critical F with 3 degrees of freedom in the numerator and 16 degrees of freedom in the denominator from Appendix table 10, we reject the null hypothesis. In the context of this problem, the average number of broken eggs is not the same for each shipping method.

In examining table 9.4, we notice that the total $SS = SST + SSE$ and that total degrees of freedom = treatment degrees of freedom + error degrees of freedom. Both of these relationships always hold in analysis of variance and provide an easy means of checking for errors in the calculations. Also, the degrees of freedom displayed in the ANOVA table are appropriate for use in looking up the critical F value from Appendix table 10.

TABLE 9.4 ANOVA Table for Egg-Shipping Method Example

Source of Variation	Degrees of Freedom	Sum of Squares	Mean Square	F
Treatments	4 − 1 = 3	89.2	89.2/3 = 29.73	14.16
Error	4(5 − 1) = 16	33.6	33.6/16 = 2.1	
Total	4(5) − 1 = 19	122.8		

Computational Formulas

While the formulas provided for treatment, error, and total sums of squares in the previous sections work, we generally do not recommend them for computing these values when we must use a hand-held calculator. A better way to compute an F test is to use a microcomputer and a spreadsheet program such as Excel or a stand-alone statistical package such as SPSS for the personal computer. Nevertheless, we present the following computational equations for use on a handheld calculator. The total sum of squares is best computed using equations 9.8 and 9.9.

$$\text{Total } SS = \Sigma\Sigma X_{ij}^2 - C \quad \boxed{9.8}$$

$$C = (\Sigma\Sigma X_{ij})^2/kn \quad \boxed{9.9}$$

The treatment sum of squares computational formula is given in equation 9.10

$$SST = \Sigma T_{.j}^2/n - C \quad \boxed{9.10}$$

where

$T_{.j}^2$ is the treatment (column) totals squared.

The error sum of squares is computed by subtraction as indicated in equation 9.11.

$$SSE = \text{Total } SS - SST. \quad \boxed{9.11}$$

We can use the data from the egg-shipping example to show how the computational formulas work:

$$C = (\Sigma\Sigma X_{ij})^2/kn = (4 + 3 + \ldots + 8)^2/(4)(5) = (108)^2/20 = 583.2$$
$$\text{Total } SS = \Sigma\Sigma X_{ij}^2 - C = (4^2 + 3^2 + \ldots + 8^2) - 583.2 = 706 - 583.2 = 122.8$$
$$SST = \Sigma T_{.j}^2/n - C = (15^2 + 38^2 + 18^2 + 37^2)/5 - 583.2 =$$
$$SST = (225 + 1444 + 324 + 1369)/5 - 583.2 = 672.4 - 583.2 = 89.2$$
$$SSE = \text{Total } SS - SST = 122.8 - 89.2 = 33.6$$

Once computed, we can place these values for sums of squares in the appropriate column of an ANOVA table, such as the third column of table 9.4, and complete the analysis as before. We remember that sums of squares cannot be negative. Thus, when using the computational formulas, we have made a mistake if we get negative values as the result of any of the subtractions.

Unequal Sample Size

While we would like to have equal-sized samples for each treatment in the analysis of variance, sometimes we cannot. For example, animals used as experimental units in an analysis of variance project may become ill and die; or weather, disease, or insects may damage experimental field plots such that some observations are lost; or for other reasons, there may not be enough data for some treatments to allow equal-sized samples. In such cases, we can still perform the analysis of variance, but the formulas for degrees of freedom and for sums of squares are somewhat different.

Sum of Squares, Degrees of Freedom, and F

The treatment sum of squares is defined as follows in equation 9.12.

$$\text{Treatment Sum of Squares} = SST = \Sigma n_j\,(\overline{X}_{.j} - \overline{X}_{..})^2 \quad \textbf{9.12}$$

In this case, we multiply the size of the jth treatment sample by the squared deviation of the jth treatment mean from the grand mean before we sum over the k treatments.

We compute the error sum of squares and total sum of squares as indicated in equations 9.13 and 9.14.

$$\text{Error Sum of Squares} = SSE = \Sigma\Sigma(X_{ij} - \overline{X}_{.j})^2 \quad \textbf{9.13}$$

$$\text{Total Sum of Squares} = \text{Total } SS = \Sigma\Sigma(X_{ij} - \overline{X}_{..})^2 \quad \textbf{9.14}$$

We obtain the degrees of freedom for treatments by the formula given in equation 9.15.

$$\text{Treatment } df = k - 1 \quad \textbf{9.15}$$

To compute error degrees of freedom, we use equation 9.16, which takes into account the different sample sizes. Also, the total degrees of freedom formula is changed to that shown in equation 9.17.

$$\text{Error } df = \Sigma(n_j - 1). \quad \textbf{9.16}$$

$$\text{Total } df = (\Sigma n_j) - 1. \quad \textbf{9.17}$$

The ANOVA table used to calculate the F statistic for one-way analysis of variance with unequal sample sizes is shown in table 9.5.

The computational formulas for the one-way analysis of variance with unequal sample sizes are also somewhat different. We will provide these as we illustrate the F test for an example problem.

TABLE 9.5 ANOVA Table for One-Way Analysis of Variance with Unequal Sample Sizes

Source of Variation	Degrees of Freedom	Sum of Squares	Mean Square	F
Treatments	$k - 1$	$SST = \Sigma n_j\,(\overline{X}_{.j} - \overline{X}_{..})^2$	$MST = SST/(k - 1)$	MST/MSE
Error	$\Sigma(n_j - 1)$	$SSE = \Sigma\Sigma(X_{ij} - \overline{X}_{.j})^2$	$MSE = SSE/\Sigma(n_j - 1)$	
Total	$(\Sigma n_j) - 1$	$SS = \Sigma\Sigma(X_{ij} - \overline{X}_{..})^2$		

Computational Formulas

Consider the following example. A production manager in an agricultural machinery plant wishes to compare the output of three stamping machines. The manager conducts an experiment with the output from each machine as a treatment by providing identical running conditions and recording the output per minute. This is done for 5 minutes for Machine A, 10 minutes for Machine B, and 6 minutes for Machine C. The data are given in table 9.6. Test at the 5 percent level of significance to determine if there is a difference in average output for the three machines. The null and alternative hypotheses for the test are:

$$H_o\colon \mu_A = \mu_B = \mu_C$$
$$H_a\colon \mu_A \neq \mu_B \neq \mu_C$$

To set up the F test for this experiment, we first calculate the sums of squares, and for this we use the computational equations depicted in equations 9.18 through 9.21. Using the data from table 9.6,

$$\text{Total } SS = \Sigma\Sigma X_{ij}^2 - C \tag{9.18}$$
$$C = (\Sigma\Sigma X_{ij})^2/\Sigma n_j \tag{9.19}$$
$$SST = \Sigma(T_{.j}^2/n_j) - C \tag{9.20}$$
$$SSE = \text{Total } SS - SST \tag{9.21}$$

we compute C first and then the total sum of squares:

$$C = (\Sigma\Sigma X_{ij})^2/\Sigma n_j = 173^2/21 = 1425.19.$$

TABLE 9.6 Production Output for Three Stamping Machines

Machine A	Machine B	Machine C
10	6	11
6	7	8
8	9	13
12	4	10
6	6	10
	10	12
	5	
	6	
	8	
	6	
$\bar{X}_{.1} = 8.4$	$\bar{X}_{.2} = 6.7$	$\bar{X}_{.3} = 10.7$
$T_{.1} = \Sigma X_{i1} = 42$	$T_{.2} = \Sigma X_{i2} = 67$	$T_{.3} = \Sigma X_{i3} = 64$

TABLE 9.7 ANOVA Table for Production Output Experiment

Source of Variation	Degrees of Freedom	Sums of Squares	Mean Square	F
Treatments	$3 - 1 = 2$	59.18	$59.18/2 = 29.59$	7.33
Error	$\Sigma(n_j - 1) = 18$	72.63	$72.63/18 = 4.035$	
Total	$(\Sigma n_j) - 1 = 20$	131.81		

$$\text{Total } SS = \Sigma\Sigma X_{ij}^2 - C = (10^2 + 6^2 + ... + 12^2) - 1425.19$$
$$\text{Total } SS = 1557 - 1425.19 = 131.81$$

To compute treatment sum of squares, we get:

$$SST = \Sigma(T_{.j}^2/n_j) - C = 42^2/5 + 67^2/10 + 64^2/6 - 1425.19$$
$$SST = 352.8 + 448.9 + 682.67 - 1425.19 = 59.18$$

Then, we obtain error sum of squares by subtraction:

$$SSE = \text{Total } SS - SST = 131.81 - 59.18 = 72.63$$

We now have the information necessary to construct the ANOVA table (table 9.7). We find the critical F for a 5 percent level of significance with 2 degrees of freedom in the numerator and 18 in the denominator in Appendix table 10 to be 3.55. Since computed $F = 7.33$ from table 9.7 exceeds 3.55, we reject the null hypothesis. In the context of this problem, the manager concludes that the average production output from the three stamping machines is not equal.

Two-Way Analysis of Variance

In experimental designs in which we use two sets of treatments simultaneously, we may choose a randomized block experiment where one set is of central importance and the other accounts for some systematic source of variability within the data and thus removes it from the experimental error. Or we may perform a completely random experiment in which we are interested in variation in both factors. In this case, there are two sets of null and alternative hypotheses, and we conduct two F tests—one for each set of treatments or factors.

Randomized Block Design

The randomized block design is a generalization of the design used for a paired t test. In that test, we paired the data and worked with differences to remove a known source of variation within the data. In the randomized block design, we

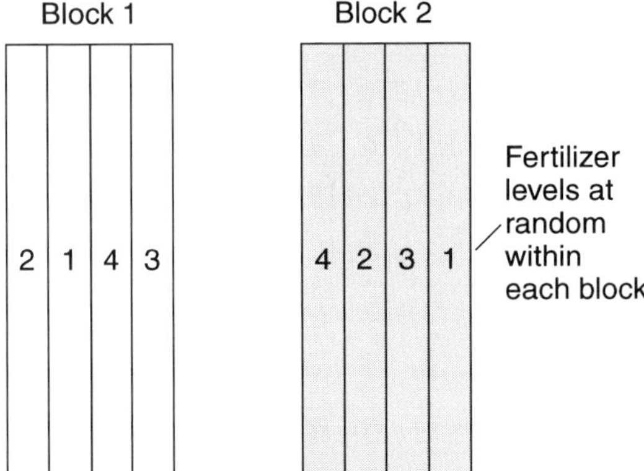

FIGURE 9.2 Field map of fertilizer equipment

create **blocks** that represent time, location, or experimental material. Thus, if we are to compare three treatments and we suspect a trend in the average response over time, we can remove a substantial portion of the time-trend variation by blocking. We do this by randomly applying the treatments to the experimental units in one small block of time and repeating the procedure in succeeding blocks of time until we collect sufficient data. Also, blocks may be parcels of land, types of animals, or some other factor in the analysis that accounts for part of the variation in the total sum of squares and may thus be removed from the experimental error by calculating a block sum of squares. For example, if an agronomist wishes to determine the effect of three levels of fertilizer application, plus a check plot with no fertilizer applied, on cotton yields using plots from two experimental farms, he would choose the farm locations as blocks to remove the effect of their different soil characteristics on yields from the error variability. Thus, his experimental design might look like the field map in figure 9.2, in which the four treatments are randomly placed at each location. Each block must contain all of the treatments or we have an incomplete block design, which is more difficult to analyze (and is beyond our scope). Also, in this design, each treatment appears only once in each block. There are randomized block designs in which the treatments are repeated or replicated within each block, but they are also beyond our scope.

The Analysis of Variance for a Randomized Block Design

The randomized block design has two qualitative independent variables—blocks and treatments. Thus, we partition the total sum of squares into three components: treatments, blocks, and experimental error. Let $\overline{X}_{i.}$ be the ith block mean and $T_{i.}$ the ith block total. Then, for a randomized block design receiving n blocks and k treatments, we have the following relationships as indicated in equations 9.22 through 9.30.

$$\text{Total } SS = SSB + SST + SSE \tag{9.22}$$

$$\text{Total } SS = \Sigma\Sigma(X_{ij} - \overline{X}_{..})^2 \text{ or} \tag{9.23}$$

$$\text{Total } SS = \Sigma\Sigma X_{ij}^2 - C \tag{9.24}$$

$$SSB = k\Sigma(\overline{X}_{i.} - \overline{X}_{..})^2 \text{ or} \tag{9.25}$$

$$SSB = \Sigma T_{i.}^2/k - C \tag{9.26}$$

$$SST = n\Sigma(\overline{X}_{.j} - \overline{X}_{..})^2 \text{ or} \tag{9.27}$$

$$SST = \Sigma T_{.j}^2/n - C \tag{9.28}$$

$$SSE = \text{Total } SS - SSB - SST \tag{9.29}$$

$$\text{where } C = (\Sigma\Sigma X_{ij})^2/kn \tag{9.30}$$

We illustrate the analysis of variance for the randomized block design in table 9.8. The second column contains the degrees of freedom associated with each sum of squares. We compute the mean squares by dividing the sums of squares by their respective degrees of freedom. There is only one computed F statistic for this test—treatment mean square divided by error mean square—because the null hypothesis being tested is that the treatment means are equal. We use blocking in this case to reduce experimental error.

Consider the following example. A plant breeder develops a new wheat variety that she hopes will yield more in the plains than the three most popular varieties under dry-land farming conditions. She sets up a randomized block experiment on the three major types of soils in the region—sand, clay, and loam. She selects a field with each soil type, divides it into four parcels, and randomly selects one of the four wheat varieties for planting on each parcel. The yields per acre from each parcel are given in table 9.9. Test to see if the average yields for each variety are equal using a 5 percent level of significance.

We first write the null and alternative hypotheses as

$$H_o: \mu_A = \mu_B = \mu_C = \mu_D$$
$$H_a: \mu_A \neq \mu_B \neq \mu_C \neq \mu_D$$

TABLE 9.8 ANOVA Table for Randomized Block Design

Source of Variation	Degrees of Freedom	Sum of Squares	Mean Square	F
Treatments	$k - 1$	SST	$MST = SST/(k - 1)$	MST/MSE
Blocks	$n - 1$	SSB	$MSB = SSB/(n - 1)$	
Error	$(k - 1)(n - 1)$	SSE	$MSE = SSE/(k - 1)(n - 1)$	
Total	$kn - 1$	Total SS		

TABLE 9.9 Dry-Land Wheat Yield Data by Variety and Soil Type

	Variety				
Block	***A***	***B***	***C***	***D***	**Totals**
Sand	20	21	21	18	80
Clay	25	24	21	20	90
Loam	30	28	22	20	100
Totals	75	73	64	58	270

TABLE 9.10 ANOVA Table for Dry-Land Wheat Variety Test on Specified Soil Types

Source of Variation	Degrees of Freedom	Sum of Squares	Mean Square	*F*
Treatments	$(4 - 1) = 3$	63	21	4.50
Blocks	$(3 - 1) = 2$	50	25	
Error	$(4 - 1)(3 - 1) = 6$	28	4.67	
Total	$(4)(3) - 1 = 11$	141		

Next, we compute the sums of squares

$$C = (\Sigma\Sigma X_{ij})^2/kn = (270)^2/(4)(3) = 6{,}075$$
$$\text{Total } SS = \Sigma\Sigma X_{ij}^2 - C = 20^2 + 25^2 + \ldots + 20^2 - 6{,}075 = 141$$
$$SST = \Sigma T_{.j}^2/n - C = (75^2 + 73^2 + 64^2 + 58^2)/3 - 6{,}075 = 63$$
$$SSB = \Sigma T_{i.}^2/k - C = (80^2 + 90^2 + 100^2)/4 - 6{,}075 = 50$$
$$SSE = \text{Total } SS - SST - SSB = 141 - 63 - 50 = 28$$

and place them in an ANOVA table to complete the analysis (table 9.10). The computed $F = 4.50$ is the ratio TMS/EMS and we compare it with the critical $F = 4.76$ from Appendix table 10 for a 5 percent level of significance with 3 degrees of freedom in the numerator and 6 degrees of freedom in the denominator. Because computed F is smaller than the critical F, we cannot reject the null hypothesis. In the context of this problem, the average dry-land wheat yields from the four varieties (including the new one) are the same.

The Completely Randomized Design

The completely randomized design is a second type of two-way analysis of variance in which we are interested in both of the qualitative variables—treatments

TABLE 9.11 Height (Inches) of Poinsettias in 8-inch Pots Treated with Growth Retardant and Fertilizer

	Level of Growth-Retardant Hormone			
Fertilizer Rate	**Check**	**Dose 1**	**Dose 2**	**Totals**
Check	24	16	14	54
Low	26	17	14	57
Medium	27	17	15	59
High	28	18	16	62
Totals	105	68	59	232

and blocks. There are two null hypotheses, one for treatments and another for blocks, which we test against their respective alternative hypotheses. Thus, there are two computed F values—one for each test—and we reject or not reject the null hypotheses based on the individual test outcomes. The formulas for degrees of freedom, sums of squares, and mean squares are identical to those for the randomized complete block experiment.

Consider the following example. An ornamental horticulturist wishes to test the response of a single variety of poinsettia plants to two levels of a growth-retardant hormone and to three rates of fertilizer application. He also includes a check for each treatment in which the plants receive none of the material; thus, we have three treatments and four blocks. After three weeks of growth in the greenhouse, he measures the height of the plants, which are in 8-inch pots. The data are presented in table 9.11. Test at the 5 percent level of significance to see if average height of the poinsettias is affected by the growth-retardant hormone and by fertilizer application rate.

The null and alternative hypotheses for this test are:

$$H_o\colon \mu_0 = \mu_1 = \mu_2 \qquad H_o\colon \mu_0 = \mu_L = \mu_M = \mu_H$$

$$H_a\colon \mu_0 \neq \mu_1 \neq \mu_2 \qquad H_a\colon \mu_0 \neq \mu_L \neq \mu_M \neq \mu_{H.}$$

To perform the analysis of variance for this problem, we first calculate the sums of squares. We compute the correction term, C as:

$$C = (\Sigma\Sigma X_{ij})^2/kn = (232)^2/(3)(4) = 4{,}485.33$$

$$\text{Total } SS = \Sigma\Sigma X_{ij}^2 - C = (24^2 + 26^2 + \ldots + 16^2) - 4{,}485.33 = 310.67$$

$$SST = \Sigma T_{.j}^2/n - C = (105^2 + 68^2 + 59^2)/4 - 4{,}485.33 = 297.17$$

$$SSB = \Sigma T_{i.}^2/k - C = (54^2 + 57^2 + 59^2 + 62^2)/3 - 4{,}485.33 = 11.33$$

$$SSE = \text{Total } SS - SST - SSB = 310.67 - 297.17 - 11.33 = 2.17$$

We can place these sums of squares in an ANOVA table to compute the F values (table 9.12). We compare the computed $F = 410.8$ for the growth-retardant hor-

TABLE 9.12 ANOVA Table for Poinsettia Experiment

Source of Variation	Degrees of Freedom	Sum of Squares	Mean Square	F
Treatments	$(3 - 1) = 2$	297.17	148.58	410.8
Blocks	$(4 - 1) = 3$	11.33	3.78	10.45
Error	$(3 - 1)(4 - 1) = 6$	2.17	0.36	
Total	$(4)(3) - 1 = 11$	310.67		

mone treatment to critical $F = 5.14$ for a 5 percent level of significance and 2 degrees of freedom in the numerator and 6 degrees of freedom in the denominator from Appendix table 10, to test the first set of hypotheses. Since computed F is much larger than the critical F, we reject the null hypothesis. The second test reveals a critical $F = 4.76$ for 3 and 6 degrees of freedom, and when we compare it to computed $F = 10.45$, we reject the null hypothesis. Thus, in the context of this problem, the average height of the poinsettia plants is different for doses of growth-retardant hormone, and is also different for the various rates of fertilizer application.

The Latin Square Design

There are cases in which we must analyze three or more treatments because of the research problem we are facing. The analysis of variance design for such problems is broadly referred to as a factorial design. The simplest of these is a Latin square, which deals with three factors on the basis of relatively few observations. We perform the experiment by arranging the levels of one factor denoted by the letters $A, B, \ldots$, etc. into an array or a matrix so that every letter appears once, and only once, in every row and column. For example, a Latin square with four levels of the variable looks like the following (see Appendix table 11 for other sizes):

$$\begin{array}{cccc} D & B & A & C \\ B & A & C & D \\ A & C & D & B \\ C & D & B & A \end{array}$$

There are two blocking factors, one for rows and another for columns, and the treatments are randomized among them. This design is useful for agricultural economists in doing consumer studies on which rows might be grocery store locations in a large city and columns might be days of the week. We know from past research that consumers buy differently on different days of the week, and that they tend to shop in stores not too far from their neighborhood. Since people live in neighborhoods according to their income level, store location is a reasonable proxy for consumer income. The treatment might be price levels for

Texas oranges, and the number sold at each price might be the data collected. Alternatively, a soil scientist might employ a Latin square design to determine soybean yield response to fertilizer treatments, and block on soil permeability and on percent slope within the field.

The hypotheses to be tested are:

$$H_o\colon \mu_A = \mu_B = \ldots = \mu_K$$

for the K treatments and

$$H_a\colon \mu_A \neq \mu_B \neq \ldots \neq \mu_K.$$

The number of treatments determines the size of the Latin square. The data are represented by $X_{ij(k)}$ and are r^2 in number for a Latin square of r rows and r columns. The $X_{ij(k)}$ variable has the first subscript for rows, the second for columns, and the third for treatments.

The Analysis of Variance for a Latin Square

The computational formulas for the Latin square for the sums of squares are presented in equations 9.31 through 9.36. After total sum of squares, formulas are for treatments (SST), rows (SSR), columns (SSC), and error (SSE). The ANOVA table is given in table 9.13.

$$C = (\Sigma\Sigma X_{ij(k)})^2/r^2 \quad \boxed{9.31}$$

$$\text{Total } SS = \Sigma\Sigma X_{ij(k)}{}^2 - C \quad \boxed{9.32}$$

$$SST = \Sigma T_k{}^2/r - C \quad \boxed{9.33}$$

$$SSR = \Sigma T_{i.}{}^2/r - C \quad \boxed{9.34}$$

$$SSC = \Sigma T_{.j}{}^2/r - C \quad \boxed{9.35}$$

$$SSE = \text{Total } SS - SST - SSR - SSC \quad \boxed{9.36}$$

We compare the computed F from table 9.13 to the critical F_α for $(r - 1)$ degrees of freedom in the numerator and $(r - 1)(r - 2)$ degrees of freedom in the

TABLE 9.13 ANOVA Table for Latin Square Design

Source of Variation	Degrees of Freedom	Sum of Squares	Mean Square	F
Treatments	$r - 1$	SST	$MST = SST/(r - 1)$	MST/MSE
Rows	$r - 1$	SSR	$MSR = SSR/(r - 1)$	
Columns	$r - 1$	SSC	$MSC = SSC/(r - 1)$	
Error	$(r - 1)(r - 2)$	SSE	$MSE = SSE/(r - 1)(r - 2)$	
Total	$r^2 - 1$	Total SS		

TABLE 9.14 Latin Square Design for 120-day Beef Ration Experiment

	Column		
Row	**I (0 %)**	**II (25 %)**	**III (50+ %)**
1 (Heifer)	*B* 150	*A* 28	*C* 256
2 (Bull)	*C* 288	*B* 148	*A* 24
3 (Steer)	*A* 60	*C* 288	*B* 156

TABLE 9.15 Row, Column, and Treatment Totals for the 120-day Feed Ration Experiment

	Column				
Row	**I (0 %)**	**II (25 %)**	**III (50+ %)**	**Row Total**	**Treatment Total**
1 (Heifer)	*B* 150	*A* 28	*C* 256	434	*A* 112
2 (Bull)	*C* 288	*B* 148	*A* 24	460	*B* 454
3 (Steer)	*A* 60	*C* 288	*B* 156	504	*C* 832
Total	498	464	436	1,398	

denominator from Appendix table 10. We reject the null hypothesis when computed F exceeds the critical F value.

As an example, consider a beef-feeding experiment in which three treatments are applied to a pen of 550-pound animals with rows equal to sex (heifer, bull, steer) and columns equal to percent brahama influence (0, 25, 50+). The treatments consist of rations that are 1.5 (A), 2.5 (B), and 3.5 (C) percent of body weight and the variable measured is pounds of gain over a 120-day period. We present the data in table 9.14. Test at the 5 percent level of significance to see if the average pounds of gain are equal for the animals on the three rations.

The hypotheses tested for this problem are:

$$H_o\colon \mu_A = \mu_B = \mu_C$$
$$H_a\colon \mu_A \neq \mu_B \neq \mu_C$$

To perform the analysis of variance, we must first total the rows, columns, and treatments. We obtain the treatment totals by summing the three values for A in table 9.14 to obtain T_A, and so on for the other treatments. See the last column of table 9.15 for the treatment totals.

To perform the analysis of variance, we first compute the correction term C:

$$C = (\Sigma\Sigma X_{ij(k)})^2/r^2 = 1{,}398^2/3^2 = 217{,}156$$

TABLE 9.16 ANOVA Table for 120-day Beef Feed Ration Experiment

Source of Variation	Degrees of Freedom	Sum of Squares	Mean Square	F
Treatments	$3 - 1 = 2$	86,472.00	86,472/2 = 43,236	4,629
Rows	$3 - 1 = 2$	834.66	834.66/2 = 417.33	
Columns	$3 - 1 = 2$	642.66	642.66/2 = 321.33	
Error	$(3 - 1)(3 - 2) = 2$	18.68	18.68/2 = 9.34	
Total	$3^2 - 1 = 8$	87,968.00		

The total sum of squares is: $\text{Total } SS = \Sigma\Sigma X_{ij(k)}{}^2 - C = 150^2 + 288^2 + \ldots + 156^2 - 217{,}156$

$\text{Total } SS = 87{,}968$

and $SST = \Sigma T_k{}^2/r - C = (112^2 + 454^2 + 832^2)/3 - 217{,}156 = 86{,}472$

while $SSR = \Sigma T_{i.}{}^2/r - C = (434^2 + 460^2 + 504^2)/3 - 217{,}156 = 834.66$

and $SSC = \Sigma T_{.j}{}^2/r - C = (498^2 + 464^2 + 436^2)/3 - 217{,}156 = 642.66.$

Thus $SSE = \text{Total } SS - SST - SSR - SSC$

$SSE = 87{,}968 - 86{,}472 - 834.66 - 642.66 = 18.68$

and we present the analysis of variance in table 9.16. The computed F from table 9.16 of 4,629 is much greater than critical $F = 19.00$ with 2 degrees of freedom in the numerator and 2 degrees of freedom in the denominator from Appendix table 10, for a 5 percent level of significance. Therefore, we reject the null hypothesis. In the context of this problem, the average daily gain is not the same for the three feed rations.

Appendix to Chapter 9 Analysis of Variance with Excel

One-Way ANOVA

To use Excel to perform a one-way analysis of variance, we select that procedure from the Data Analysis submenu of the Tools Menu after we have entered the data in the spreadsheet using one column for each treatment. We enter the treatment labels in the first row so we can interpret the output easier. After selecting the one-way analysis of variance procedure, we enter the

TABLE 9.17 Spreadsheet Output for One-Way Analysis of Variance

SUMMARY

Groups	*Count*	*Sum*	*Average*	*Variance*
A	5	15	3	2.5
B	5	38	7.6	1.3
C	5	18	3.6	3.3
D	5	37	7.4	1.3

ANOVA

Source of Variation	*SS*	*df*	*MS*	*F*	*P-value*	*F crit*
Between Groups	89.2	3	29.73333	14.15873	9.12E-05	3.238867
Within Groups	33.6	16	2.1			
Total	122.8	19				

range of the data in the first box of the pop-up table. We check the labels box, confirm that alpha is 0.05, and write the output range in the box provided and click OK. We obtain the output in two parts. The numbers in the top part of table 9.17 are a data summary and the bottom part contains the analysis of variance. The program calls treatments Between Groups and error Within Groups, but the output values are the same as those we obtained earlier. The table *F* value is in the last column.

Unequal ANOVA

The Excel One-Way Analysis of Variance procedure works equally well with treatments that have unequal sample sizes and with equal sample sizes. Thus, we enter the data in the spreadsheet, select the One-Way Analysis of Variance procedure from the Data Analysis submenu of the Tools Menu, and in the pop-up table that appears, provide the needed information and click OK. For the example we did earlier, the computer results are as shown in table 9.18. Since we are using the same procedure, the format of the output is the same as that for table 9.17, with summary data in the first half of the table and the ANOVA results in the second half. Notice that computed *F* is the same as that computed earlier.

Randomized Block ANOVA

To use Excel with the randomized block design, we input the data, no totals, with row and column headings into the spreadsheet and select the ANOVA: Two-Factor Without Replication procedure from the Data Analysis submenu of

TABLE 9.18 Spreadsheet Results for One-Way Analysis of Variance with Unequal Sample Sizes

SUMMARY

Groups	*Count*	*Sum*	*Average*	*Variance*
Machine A	5	42	8.4	6.8
Machine B	10	67	6.7	3.344444
Machine C	6	64	10.66667	3.066667

ANOVA

Source of Variation	*SS*	*df*	*MS*	*F*	*P-value*	*F crit*
Between Groups	59.17619	2	29.5881	7.332525	0.004685	3.554561
Within Groups	72.63333	18	4.035185			
Total	131.8095	20				

the Tools Menu. In the pop-up table that appears, input the range for the data including the row and column headings in the first box, check the labels box, confirm that alpha is 0.05, write the output range in the last box, and click OK. The output appears in table 9.19. Notice that the summary portion of the table gives the totals, means, and variances for the rows first and then the columns. The ANOVA section of the table is constructed in the same manner. The variation due to rows is presented first. Since we used rows as blocks in this analysis and are not interested in that test, we ignore the computed and table *F* values for rows. We are interested in the column computed *F*, however, since that is the test on the yields associated with our varieties.

Completely Random Design ANOVA

This procedure is the same as the randomized block design with Excel because we use the same analysis of variance procedure. But, in this case, we are interested in both *F* tests. So for the poinsettia example, we first input the data including row and column labels into the spreadsheet, no totals, and select the ANOVA: Two-Factor Without Replication procedure from the Data Analysis submenu of the Tools Menu. In the pop-up table that appears, we enter the range of the data in the first box, check the labels box, confirm that alpha is 0.05, select an output range, and click OK. The results are in table 9.20. The computed *F* values agree with those obtained earlier except for rounding.

TABLE 9.19 Spreadsheet Results for Randomized Block Analysis of Variance

SUMMARY	*Count*	*Sum*	*Average*	*Variance*
Sand	4	80	20	2
Clay	4	90	22.5	5.666667
Loam	4	100	25	22.66667
Variety *A*	3	75	25	25
Variety *B*	3	73	24.33333	12.33333
Variety *C*	3	64	21.33333	0.333333
Variety *D*	3	58	19.33333	1.333333

ANOVA

Source of Variation	*SS*	*df*	*MS*	*F*	*P-value*	*F crit*
Rows	50	2	25	5.357143	0.0462584	5.143249
Columns	63	3	21	4.5	0.0558483	4.757055
Error	28	6	4.666667			
Total	141	11				

TABLE 9.20 Spreadsheet Results for Randomized Block with Completely Random Design

SUMMARY	*Count*	*Sum*	*Average*	*Variance*
Fertilizer check	3	54	18	28
Low	3	57	19	39
Medium	3	59	19.66667	41.33333
High	3	62	20.66667	41.33333
Hormone check	4	105	26.25	2.916667
Dose 1	4	68	17	0.666667
Dose 2	4	59	14.75	0.916667

ANOVA

Source of Variation	*SS*	*df*	*MS*	*F*	*P-value*	*F crit*
Rows	11.33333	3	3.777778	10.46154	0.0084802	4.757055
Columns	297.1667	2	148.5833	411.4615	3.792E-07	5.143249
Error	2.166667	6	0.361111			
Total	310.6667	11				

Exercises

1. The following experiment was designed to determine the effect of four different feeds on weight gain for pigs. Twenty pigs were randomly divided into four lots and each lot was fed a different feed. The following list shows the number of pounds gained on the feeds. Test at the 5 percent level of significance to see if the average weight gains are equal for each feed.

Ful-O-Vigor	**Peptone**	**Accu-Ration**	**Echo Feeds**
130	160	207	192
147	151	236	187
131	148	216	195
153	150	230	191
141	155	227	190

2. A lot of stocker calves was sorted into groups due to frame size, and weight gain after 60 days on wheat pasture is recorded in the following list. Test at the 5 percent level of significance to determine if average weight gains are equal by size of frame.

Large	**Medium**	**Small**
175	144	123
184	151	118
168	157	130
	148	125
	152	
	140	
	155	

3. An agricultural economist classified medium-sized farms that did not reach the farm program payment limit according to type and section of the country. Test to see if average program payment received (in $000s) is the same by type of farm and by region. Use $\alpha = 0.05$.

	Farm Type		
Region	**Cotton**	**Feed Grain**	**Food Grain**
Southeast	31	38	27
Central	46	42	35
West	30	26	48

4. A horticulturist sprayed mature pecan trees at various locations within the orchard with several liquid fertilizer formulations at 2-week intervals to determine yield response. The yield in pounds of pecans per tree is given in the following list. Test to see if the average yields are the same for the

various fertilizer formulations at the 0.01 level of significance using locations as blocks.

	Fertilizer with N, Zn		
Location	**High**	**Medium**	**Low**
By stream	138	120	108
Center orchard	90	78	70
In draw	115	89	76

5. Examine the following experimental design schemes and state which are Latin squares.

a. *A B C*
C A B
B C A

b. *A B C*
C C A
B A B

c. *A B C*
C A B
A C A

d. *B C D A*
A D B C
D A C B
C B A D

e. *B C D A*
D A B C
A C D B
C B A D

6. Complete the following analysis of variance table.

Source of Variation	Degrees of Freedom	Sum of Squares	Mean Square	*F*
Treatments		135.00	33.750	
Blocks	3		23.333	
Error				
Total		250.00		

7. An agricultural economist is interested in market-testing a new food product made from potatoes. She sets up a Latin square design with three treatments (prices set at A = \$0.95, B = \$1.05, and C = \$1.15) over three weeks in three stores (in three income areas of the city). Test at the 0.01 level of significance to see if average sales are the same for all three prices.

	Columns		
Rows	***I (Store 1)***	***II (Store 2)***	***III (Store 3)***
1 (Week 1)	50 (*C*)	79 (*B*)	105 (*A*)
2 (Week 2)	74 (*B*)	106 (*A*)	63 (*C*)
3 (Week 3)	98 (*A*)	61 (*C*)	93 (*B*)

8. Use Excel to do exercises 1 and 2.
9. Use Excel to do exercises 3 and 4.

CHAPTER

10

Chi-square Applications

In the previous chapter, we discussed procedures for testing hypotheses about the equality of three or more means. We now turn to the examination of tests concerning three or more population proportions. The sampling distribution employed in these tests is the chi-square, a probability distribution we used to test hypotheses about a single variance in Chapter 7. Other applications that we can make with chi-square include using sample data to test whether two variables are independent, and fitting sample data to distributions to see if, for example, the data come from a normal population or from a binomial distribution. We examine these applications in turn.

K Sample Tests for Proportions

When we select k (>2) samples from a population with the intent to test the null hypothesis of whether the proportions of successes in those samples are equal, we have a chi-square test that is one-sided to the right with $(k - 1)$ degrees of freedom. Degrees of freedom refer to the number of restrictions placed on the data. We require the number of observations in each sample to add to the total for the entire group. This is a restriction since expected frequencies depend upon the group total. Once we establish the degrees of freedom and the level of significance, we then determine the critical value of chi-square from Appendix table 9. If the value of chi-square computed from the data exceeds the critical value, we reject the null hypothesis (figure 10.1). The formula for computed chi-square is contained in equation 10.1, in which f_o is

$$\chi^2 = \Sigma \frac{(f_o - f_e)^2}{f_e} \qquad \textbf{10.1}$$

an observed sample frequency and f_e is an expected frequency if the null hypothesis is true.

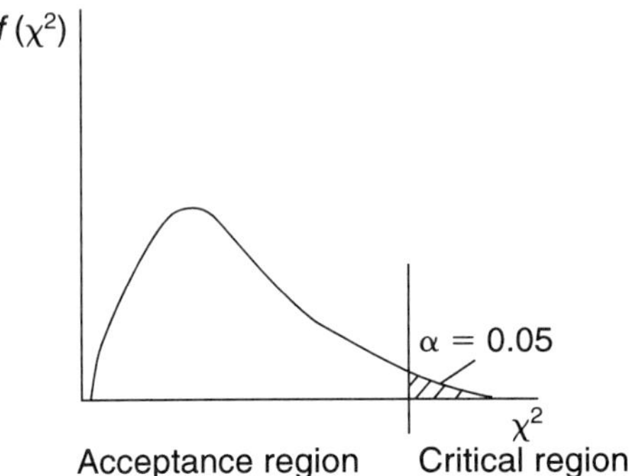

FIGURE 10.1 Acceptance and rejection regions for chi-square distribution at the 5 percent level

As an example, consider a test conducted by an agricultural machinery manufacturing company manager. He has three operators of the same machine on different shifts and records the number of defective parts produced on a recent shift by all three. He performs a three-sample test of proportions to determine if the proportion of defective parts produced by the three operators is the same (table 10.1). The hypotheses for the test are:

$$H_o: p_1 = p_2 = p_3 = 0.33$$
$$H_a: \text{the p's aren't all equal}$$

We obtain the expected number of defectives, f_e, for the computed chi-square formula displayed in column three of table 10.1 by multiplying the total number of observed frequencies (75) by the proportions stated in the null hypothesis (0.33). The total of the last column of table 10.1 is the value for computed chi-square, $\chi^2 = 1.68$. The critical value of chi-square from Appendix table 9 for $(3 - 1) = 2$ degrees of freedom and a 5 percent level of significance is $\chi^2 = 5.991$.

TABLE 10.1 Chi-square Test of Proportion of Defectives Produced by Three Machine Operators

Machine Operator	f_o, Observed Number of Defectives	f_e, Expected Number of Defectives	$f_o - f_e$	$(f_o - f_e)^2$	$(f_o - f_e)^2/f_e$
Smith	24	25	−1	1	0.04
Jones	30	25	5	25	1.00
Garza	21	25	−4	16	0.64
Total	75	75			1.68

Since the computed chi-square is less than the critical value, we cannot reject the null hypothesis. In the context of this problem, the proportion of defective parts produced by each machine operator is the same.

Tests of Independence in Contingency Tables

We often summarize data in contingency tables in which we present several levels of one variable in the columns of the table and those of a different variable down the stub (or as row headings). We are interested in testing whether the variables are independent. If they are independent, then the values of the response variable in the body of the table do not depend upon the classification scheme we used in the column and row headings. However, if the variables are not independent, then the response variable values are related to the classification. We use the chi-square distribution as the basis for this test. The computed chi-square is small when the row and column variables are independent and larger than the critical chi-square when the variables are not independent. The formula for computed chi-square is the same as in the previous section, but we compute degrees of freedom differently. In this case, we multiply the number of rows minus one by the number of columns minus one to get degrees of freedom, as indicated in equation 10.2.

$$\nu = (r - 1)(c - 1). \quad \boxed{10.2}$$

We calculate the expected values (f_e) for each table cell that are required for the computed chi-square formula as the product of the row total and the column total for that cell divided by the grand total for all of the table observations, as shown in equation 10.3. The number of expected

$$f_e = \text{(row total)(column total)/(grand total)}. \quad \boxed{10.3}$$

values that we must compute in this manner are equal to the number of degrees of freedom. We can obtain the remaining expected values by subtraction since the expected values for each table cell must add to give the appropriate row and column totals for the table.

Consider the following example. The owner of a local farm machinery repair shop surveyed past repair orders for the calendar year. On last year's tractor model only, she noted whether repairs were electrical, fuel related, or other. She categorized the data by make of tractor and wishes to test whether type of repair is independent of tractor make at the 5 percent level of significance. For data of this type, if there are fewer than five observations per cell, the chi-square test does not work properly and some table cells need to be combined. Once we have the observed data, we calculate the expected frequencies, which appear inside parentheses in table 10.2. For example, we compute the expected frequency for the cell for tractor make A and electrical repairs as

$$f_e = \text{(row total)(column total)/(grand total)} = (43)(77)/(230) = 14.4.$$

TABLE 10.2 Number of Tractors Repaired by Make and Kind of Repair

Make of Tractor	Type of Repair			
	Electrical	*Fuel Supply*	*Other*	*Total*
A	17 (14.4)	19 (18.7)	7 (9.9)	43
B	14 (10.1)	7 (13.0)	9 (6.9)	30
C	6 (13.0)	21 (17.0)	12 (9.0)	39
D	33 (32.2)	44 (41.7)	19 (22.1)	96
E	7 (7.3)	9 (9.6)	6 (5.1)	22
Total	77	100	53	230

Once we obtain all of the expected frequencies, we can then employ the computed chi-square formula to calculate a value to compare with the critical chi-square.

$$\chi^2 = \Sigma \frac{(f_o - f_e)^2}{f_e} = \frac{(17 - 14.4)^2}{14.4} + \frac{(19 - 18.7)^2}{18.7} + \ldots + \frac{(6 - 5.1)^2}{5.1} = 12.74$$

The degrees of freedom for this problem are $\nu = (r - 1)(c - 1) = (5 - 1)(3 - 1) = (4)(2) = 8$. The critical value of chi-square from Appendix table 9 for a significance level of 5 percent and 8 degrees of freedom is 15.507. Since the computed $\chi^2 = 12.74$ is less than the critical value, we do not reject the null hypothesis of independence. For this problem, tractor make is independent of type of repair, i.e., the repairs are randomly distributed among makes of tractor.

In the case of contingency tables with only two rows and two columns, there is a short computed chi-square formula, but we recommend using the personal computer rather than learning it.

Goodness-of-Fit Tests

In many statistical problems, we are interested in learning how closely an observed set of sample data fits a theoretical probability distribution. We can use the chi-square distribution as the basis for tests of this nature. The computed chi-square formula used in the earlier sections of this chapter also applies here, with the expected frequencies calculated as the total sample observations multiplied by the theoretical probabilities of the frequency distribution being tested under the null hypothesis. Degrees of freedom for the test is the number of data categories minus the restrictions placed on the data. If we must use the data to estimate values of the parameters necessary for defining the theoretical probability distribution being tested, then we lose a degree of freedom for each parameter estimate in addition to the one lost because the sample total must be calculated. For example, if we decide to fit a set of data to a normal distribution, then we must know the mean and standard deviation before we can use the Z formula to calculate the theoretical probabilities for the normal distribution. If we do not know from outside information what those values are so that we estimate them with the data, then we lose two additional degrees of freedom. Thus, we have $(k - 3)$ degrees of freedom for the goodness-of-fit test rather than $(k - 1)$. Common distributions to which we fit sets of data include the normal, binomial, Poisson, and uniform. We will consider cases that apply to the first two distributions with appropriate examples.

For the first example, consider a large grain company that has 850 hourly employees with wages distributed as in table 10.3. Test to see if the data are normally distributed. The mean and standard deviation estimated from the hourly wage data are $6.50 and $0.35, respectively.

To fit a normal distribution to the data, we first calculate Z values for each class limit for the frequency distribution in table 10.3, look up the probabilities for the calculated Z values, and compute the probability that the random variable would fall in the class, usually by subtraction. Then we write these probabilities for each class in a column of the table. Once these are obtained, we multiply each probability by the grand total for the data (850) to calculate expected frequencies for that class, and finally we compute chi-square using our observed and expected frequencies and compare it to the critical chi-square to test our hypothesis.

The null and alternative hypotheses for the test are:

H_o: The data fit a normal distribution with mean $6.50, σ = $0.35.
H_a: The data do not fit such a normal distribution.

TABLE 10.3 Hourly Wage Data for Grain Company Employees

Hourly Wage	Number of Employees
$5.75–6.00	62
$6.00–6.25	124
$6.25–6.50	267
$6.50–6.75	228
$6.75–7.00	106
$7.00–7.25	46
$7.25–7.50	12
$7.50–7.75	5
Total	850

Now we compute the first two Z values:

$$Z = \frac{X - \mu}{\sigma} = \frac{5.75 - 6.50}{0.35} = -2.14 \; P(Z = -2.14) = 0.4838$$

$$Z = \frac{X - \mu}{\sigma} = \frac{6.00 - 6.50}{0.35} = -1.43 \; P(Z = -1.43) = 0.4236$$

Subtracting the two probabilities gives the probability that a normally distributed random variable X is between \$5.75 and \$6. Thus,

$$0.4838 - 0.4236 = 0.0602$$

is the probability for this class, but if we use it, we ignore the area under the left tail of the normal distribution that is less than \$5.75. Thus, we subtract 0.4236 from 0.5000 to obtain 0.0764, which goes in the first cell of column 3 in table 10.4. To compute f_e for this class, multiply the total number of employees by the probability we just computed. Thus,

$$f_e = (850)(0.0764) = 64.9,$$

which is the value for the first cell of column 4 in table 10.4. We continue with this process until all of the expected frequencies are attained for column 4. We can obtain computed chi-square by using the following formula:

$$\chi^2 = \Sigma \frac{(f_o - f_e)^2}{f_e} = \frac{(62 - 64.9)^2}{64.9} = \frac{(124 - 138)^2}{138} + \ldots + \frac{(5 - 1.9)^2}{1.9} = 23.84$$

The degrees of freedom for this test are $(k - 3) = (8 - 3) = 5$. Critical chi-square for a significance level of 5 percent and 5 degrees of freedom from Appendix

TABLE 10.4 Theoretical Probabilities and Expected Frequencies for Hourly Wage Data

Hourly Wage	Number of Employees	Normal Probabilities	Expected Frequencies
\$5.75–6.00	62	0.0764	64.9
\$6.00–6.25	124	0.1624	138.0
\$6.25–6.50	267	0.2612	222.0
\$6.50–6.75	228	0.2612	222.0
\$6.75–7.00	106	0.1624	138.0
\$7.00–7.25	46	0.0602	51.2
\$7.25–7.50	12	0.0141	12.0
\$7.50–7.75	5	0.0021	1.9
Total	850	1.0000	850

table 9 is 11.07. Since the computed chi-square is 23.84, we reject the null hypothesis and claim that the hourly wage data in table 10.3 do not fit a normal distribution with mean \$6.50 and standard deviation \$0.35. If we were planning to use some statistical procedure on this set of data that required a normal distribution, we would not be able to do so.

We now consider our second example—a goodness-of-fit test for a binomial probability distribution. A commercial swine producer has collected data during the past year on 1,000 litters of eight pigs each with the following number of males. Do these data fit a binomial distribution with $n = 8$ and $p = 0.5$ at the 5 percent level of significance? To solve this problem, we first turn to Appendix table 2 and write down the probabilities for our binomial distribution opposite each value of r in table 10.5. To compute the values of f_e, the expected frequencies, we next multiply the total of 1,000 litters by each of the binomial probabilities and record the result in table 10.5, column 4. We can then perform the test. Our computed chi-square is calculated as 6.4 using our

$$\chi^2 = \Sigma \frac{(f_o - f_e)^2}{f_e} = \frac{(6 - 3.9)^2}{3.9} + \frac{(27 - 31.2)^2}{31.2} + \ldots + \frac{(7 - 3.9)^2}{3.9} = 6.4$$

computed chi-square formula. We compare the computed chi-square value of 6.4 to the critical chi-square of 15.507 from Appendix table 9 for 8 degrees of freedom and a 5 percent level of significance. On the basis of this comparison, we cannot reject the null hypothesis that the data fit a binomial distribution with $n = 8, p = 0.5$. Thus, the number of males in litters of eight pigs each for this grower is as expected by the genetic model, which says the proportion should be 0.5.

TABLE 10.5 Number of Males in Litters of Eight Pigs and Binomial Probabilities and Expected Frequencies for a Chi-square Goodness-of-Fit Test

Number of Male Pigs, r	Observed Number of Litters, f_o	Binomial Probabilities	Expected Number of Litters, f_e
0	6	0.0039	3.9
1	27	0.0312	31.2
2	115	0.1094	109.4
3	226	0.2188	218.8
4	271	0.2734	273.4
5	215	0.2188	218.8
6	98	0.1094	109.4
7	35	0.0312	31.2
8	7	0.0039	3.9
Total	1,000	1.0000	1,000

Exercises

1. A marketing researcher wants to know whether consumers prefer the taste of any of four different types of orange juice: fresh squeezed, frozen concentrate, reconstituted from powder, or canned. She selects a random sample of 100 orange juice drinkers. Each person is randomly given all four types served in glasses marked *A* through *D*, respectively, and is asked to state which one is preferred. The following data show the results. Test at the $\alpha = 0.05$ level to see if the proportion favoring each drink is the same.

Brand Preferred	Number of Consumers
A	33
B	29
C	21
D	17

2. Two hybrid flowers were crossed by a horticulturist with the result that petals were of the following colors: 122 purple, 50 blue, 38 red, and 14 mottled. Do these results disagree with the theoretical frequencies of a 9:3:3:1 ratio? Use a 0.01 level of significance.

3. Seed potatoes were treated chemically for blight, with the results given in the following table. Are the number of living potato plants independent of treatment received? Test at the 0.05 level of significance.

Treatment	Living	Dead
Inoculated	152	48
Not treated	33	67

4. A county agricultural extension agent just moved to a new assignment. As part of his orientation to the new job, he looked at data on the number of livestock producers in the county classified by full-time and part-time producers, and related that information to records in the office concerning participation in livestock education programs conducted by his predecessor. Is participation independent of type of producer at the 0.01 level of significance?

Type of Producer	Participant	No Contact
Full-time, large herd	17	33
Full-time, medium herd	21	14
Part-time, small herd	32	13

5. The following data represent the annual rainfall in a certain location during a 65-year period. Test at the 0.05 level to see if rainfall is normally distributed with a mean of 18 inches and standard deviation of 7.9.

Inches	*Frequency*
Less than 10	9
10–15	14
15–20	18
20–25	11
25–30	8
30 or more	5

6. Test to see if the number of farmers arriving at a gin with cotton to be unloaded during a 15-minute interval fit a Poisson distribution with a mean of 2 using a 5 percent level of significance.

Number of Farmers	*Number of 15-Minute Intervals*
0	22
1	50
2	16
3	10
4 or more	7

7. An animal scientist wants to publish her findings about the effects of a special diet on variability of litter size in swine. In her study, she randomly selected twenty-five female pigs and raised them through pregnancy on the diet. The mean and variance for this sample were $\bar{X} = 8.3$ and $S^2 = 3.4$. Construct a 90 percent confidence interval for the population variance for the litter size of pigs on this diet.

8. In a quality control department of a hide tanning plant, samples of size 5 are taken from the processing line and the number of defects recorded. After 1,000 samples had been taken, the following data were summarized. Test to see if they fit a binomial distribution with $p = 0.05$.

Number of Defects	*Number of Samples*
0	752
1	210
2	25
3 or more	13

CHAPTER 11

11

Correlation, Regression, and Time Series

Until now, we have used information contained in the data collected for a random variable X to make inferences about values of interest that X might assume. Thus, we have tested the mean of X, using the mean as a proxy for most of the values in the data set. Now we consider information contained in some other variable, Y, and test to see if a relationship exists between X and Y that will explain more about X than the mean of X will tell us. With a pair of variables, we have two possible kinds of analysis. With one, both variables are random and we are interested in the degree of association between the two. If the degree of association is strong so that if one variable increases in value and the other can also be expected to increase, and if we can only observe values for one variable in the set, we can use them to help estimate quantities of the variable we cannot observe. With the other kind of analysis, one variable is random while the other is fixed so that we can know with certainty the values it will assume. And, if we can mathematically describe their relationship, we have a way of estimating the value that the random variable is expected to assume when the fixed variable takes on a given value. The first kind of analysis is correlation, while the second is regression. We take up these in turn.

Correlation Analysis

Two random variables, X and Y, are correlated depending upon whether they move in the same or in opposite directions, or in no observable pattern. In the latter case, the variables are not correlated, but independent. For example, the price of wheat is independent of the price of desktop computers because completely different market forces determine their values. If the two variables happen to move together for a short while, we would consider it a short-term, random event. However, the prices of wheat and barley generally tend to move in the same direction because both are affected by similar market forces, and are positively correlated. Likewise, the price and quantity of wheat consumed are random variables that move in opposite directions and are negatively correlated. We can think of many other examples of correlated and uncorrelated random variables.

The **correlation coefficient,** r, measures the strength of the degree of association between two random variables. It ranges in value from ± 1, i.e., $-1 \leq r \leq 1$. For independent random variables, the value of r is close to zero, while it is positive for directly related variables and negative for those that are inversely related. The higher the absolute value of r, the closer the degree of association. We present two computational formulas for r as stated in equations 11.1 and 11.2.

$$r = \frac{\Sigma XY - \dfrac{\Sigma X \Sigma Y}{n}}{\sqrt{\left[\Sigma X^2 - \dfrac{(\Sigma X)^2}{n}\right]\left[\Sigma Y^2 - \dfrac{(\Sigma Y)^2}{n}\right]}} \qquad \textbf{11.1}$$

$$r = \frac{\Sigma XY - n\overline{X}\overline{Y}}{\sqrt{(\Sigma X^2 - n\overline{X}^2)(\Sigma Y^2 - n\overline{Y}^2)}} \qquad \textbf{11.2}$$

Either is appropriate. Because we always take the positive square root from the denominator of these two equations, r is negative when the result of the subtraction performed in the numerator is negative.

We can use **scatter diagrams** to display the **correlation** between the random variables X and Y graphically. In a scatter diagram, we plot each (X_i, Y_i) pair for the data as a point and examine the resulting "scatter" (figure 11.1). All of the points fall on a straight line if $r = 1$, which is positively sloped when r is positive and negatively sloped when r is negative. The points fall in a narrow band or ellipse when r is less than 1, which becomes more like a line as the r values approach one. When r is zero, the points fall in a circle or a mass with no apparent direction. Small values of r produce a scatter diagram with points that are more like a circle than a line.

We may test the correlation coefficient with the t distribution. The null and alternative hypotheses are:

$$H_o\colon \rho = 0$$
$$H_a\colon \rho < 0 \text{ (or } \rho > 0)$$

where ρ (the Greek letter rho) represents the population correlation coefficient. Degrees of freedom for the test are $(n - 2)$ because we lose a degree of freedom for estimating the means of both X and Y, and n represents the number of pairs

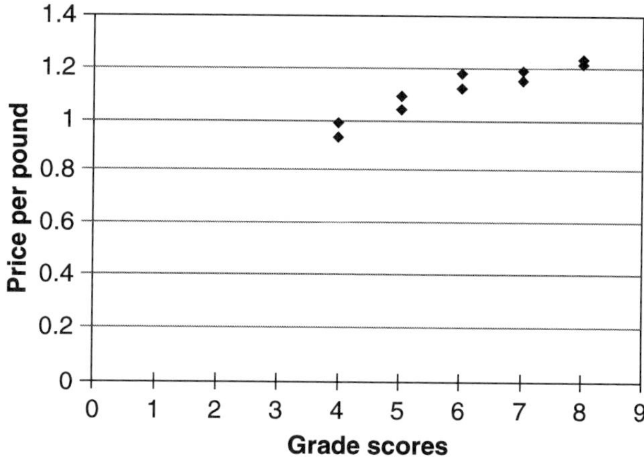

FIGURE 11.1 Scatter diagram of Angus carcass data

of data in our set. We compute t using equation 11.3, and since $\rho = 0$ under the null hypothesis, we usually leave it out of the numerator of computed t.

$$t = \frac{r - \rho}{S_r} = \frac{r - \rho}{\sqrt{\dfrac{1 - r^2}{n - 2}}} \quad \boxed{11.3}$$

Consider the following example. Beef carcass quality grades are designated by the U.S. Department of Agriculture as follows: Low Prime (slightly abundant marbling; score = 8), High Choice (moderate marbling; score = 7), Average Choice (modest marbling; score = 6), Low Choice (small amount of marbling; score = 5), Select (slight marbling; score = 4), High Standard (traces of marbling; score = 3), and Average Standard (practically devoid of marbling; score = 2). An animal scientist wishes to determine the degree of association between carcass quality grade scores assigned by a U.S. Department of Agriculture grader (X) and the selling price per pound of the carcass (Y) for a random sample of ten Angus steer carcasses. The data are in table 11.1. We calculate r by placing the values from the table into one of the computational equations and solving. We obtain $r = 0.94$,

$$r = \frac{\Sigma XY - \dfrac{\Sigma X \Sigma Y}{n}}{\sqrt{\left[\Sigma X^2 - \dfrac{(\Sigma X)^2}{n}\right]\left[\Sigma Y^2 - \dfrac{(\Sigma Y)^2}{n}\right]}} = \frac{68.07 - \dfrac{(60)(11.14)}{10}}{\sqrt{\left[380 - \dfrac{(60)^2}{10}\right]\left[12.495 - \dfrac{(11.14)^2}{10}\right]}} = 0.94$$

which indicates that the degree of association between carcass quality grade scores and price received per pound by the producer is quite good. We can test

TABLE 11.1 Beef Carcass Quality Grade Scores and Price Received Per Pound for Angus Steer Carcasses

Quality Grade Score, X	Selling Price per Pound, Y	XY	X^2	Y^2
5	$1.09	5.45	25	1.188
7	1.19	8.33	49	1.416
4	0.99	3.96	16	0.980
5	1.04	5.20	25	1.082
8	1.21	9.68	64	1.464
6	1.18	7.08	36	1.392
4	0.94	3.76	16	0.884
7	1.15	8.05	49	1.322
8	1.23	9.84	64	1.513
6	1.12	6.72	36	1.254
60	11.14	68.07	380	12.495

this value to see if it is indeed greater than zero by using the t test. Our hypotheses are:

$$H_o: \rho = 0$$
$$H_a: \rho > 0.$$

We employ a one-sided test to the right for this problem because we expect a positive correlation between the two random variables—carcass quality scores and price received. The degrees of freedom for the problem are

$$\nu = (n - 2) = (10 - 2) = 8.$$

If we choose a 5 percent level of significance for the test, then the critical t from Appendix table 8 is 1.860. If computed t is greater than 1.860, we must reject the null hypothesis; otherwise not. Computed t is calculated using our formula as $t = 7.79$.

$$t = \frac{r - p}{\sqrt{\frac{1 - r^2}{n - 2}}} = \frac{0.94 - 0}{\sqrt{\frac{1 - 0.94^2}{10 - 2}}} = 7.79$$

Because computed $t = 7.79$ is greater than the critical $t = 1.860$, we reject the null hypothesis of no correlation between the two random variables—carcass quality grade scores and price received per pound. The two variables tend to move in the same direction for Angus steer carcasses (figure 11.1).

Simple Linear Regression

Regression analysis is designed to indicate a mathematical relationship between a fixed variable X and a random variable Y in an equation that makes Y the dependent variable. Thus, we have $Y = a + bX$ as the type of equation. This equation is linear in its parameters, a and b, where a must be the intercept term and b the slope if we are to have an equation that describes a straight line. To know the form of the equation, we must collect pairs of observations (X_i, Y_i) and use them to obtain estimates of the equation parameters, a and b. For example, the line $Y = 42 - 2X$ is different from the line $Y = -100 + 0.7X$, and their differences are completely described by the values of a and b. Mathematically, we may use a number of methods to obtain values for a and b, given a set of observations on X and Y. However, the least squares procedure, which focuses on the random variable Y, is used for regression analysis. Specifically, the **method of least squares** minimizes the sum of squared deviations of each data point (X_i, Y_i) from the line being estimated in the Y direction. Since we have no reason to expect all of the raw data points (X_i, Y_i) to fall on a straight line, for each value of X_i in the data, there may be two values of Y: the original data point, Y_i, and the value from the estimated equation, Y_e, that depends on a and b. The difference $(Y_i - Y_e)$ is the deviation that we are discussing. Since $\Sigma(Y_i - Y_e) = 0$, it is not of much use to us; however the $\Sigma(Y_i - Y_e)^2$ is not zero and we may select it as the criterion for fitting a line to the data set. Therefore, the method of least squares minimizes $\Sigma(Y_i - Y_e)^2$. Mathematically, we minimize this sum of squares by taking its derivatives with respect to a and b in turn and setting them equal to zero and solving. This results in the **Normal Equations** (equation 11.4). We can solve the

$$\begin{aligned}\Sigma Y_i &= na + b\Sigma X_i \\ \Sigma X_i Y_i &= a\Sigma X_i + b\Sigma X_i^2\end{aligned} \tag{11.4}$$

first equation for a to give the definitional and computational formulas (equations 11.5 and 11.6).

$$a = \overline{Y} - b\overline{X} \tag{11.5}$$

$$a = 1/n(\Sigma Y_i - b\Sigma X_i) \tag{11.6}$$

By substituting the right-hand side of this last equation into the second normal equation, we also get two expressions for b (equation 11.7). The equation involving the means of X and Y is

$$b = \frac{\Sigma(X_i - \overline{X})(Y_i - \overline{Y})}{\Sigma(X_i - \overline{X})^2} = \frac{\Sigma X_i Y_i - \dfrac{(\Sigma X_i)(\Sigma Y_i)}{n}}{\Sigma X_i^2 - \dfrac{(\Sigma X_i)^2}{n}} \tag{11.7}$$

easy to use with a handheld calculator if both means are whole numbers; otherwise we find the second expression for b easier to use. Regression analysis is best done with a microcomputer program such as Excel or SPSS for the personal computer.

The preceding equations for a and b are the direct result of the least squares estimation procedure, which provides the best[1] estimators because they are unbiased and more precise than any other set. Thus, we want to use these formulas when performing regression analysis to obtain the best equation possible for explaining a linear relationship between X and Y.

Once we have used the least squares formulas for a and b and have the equation for the linear relationship between X and Y, how do we know that the equation provides more information about Y than the mean of Y by itself, ignoring X? We can find out by testing hypotheses about the slope parameter, b, in the least squares linear equation. If the slope, b, is actually zero, then there is no linear relationship between X and Y—there is just a horizontal line with a height equal to the mean of Y since the second term in the least squares formula for a becomes zero when $b = 0$. This is equivalent to saying that the mean of Y is the best estimator of Y. On the other hand, if b is not zero, then the fixed variable X adds more to the explanation of the variability in Y than the mean of Y by itself; hence our test on b. In simple linear regression like this where there is only one X variable, and one b, either of two tests may be performed—a t test on b or an analysis of variance on the linear equation. We will examine these in turn.

A *t* Test on *b*

We state the null and alternative hypotheses for this test in terms of the population parameter for the slope of the line, β. The alternative hypothesis can be two-sided if we have no external information to indicate the direction of the relationship between X and Y, or one-sided if we have data that provide a direction (positive or negative) for the relationship. We write a two-sided test as:

$$H_o: \beta = 0$$
$$H_a: \beta \neq 0$$

We get degrees of freedom for this test by equation 11.8 in which k equals the number of X

$$\nu = (n - k - 1) \qquad \boxed{11.8}$$

variables in the equation. We lose a degree of freedom for estimating a and another for each b we estimate. In the case of simple linear regression, degrees of freedom are $(n - 2)$ since there is one a and one b. Once we know the level of significance, we can obtain critical t from Appendix table 8 and compare it with computed t as expressed in equation 11.9. If computed t exceeds the critical

$$t = \frac{b - \beta}{S_b} = \frac{b - \beta}{\sqrt{\dfrac{\dfrac{\Sigma(Y_i - Y_e)^2}{n - 2}}{\Sigma(X_i - \overline{X})^2}}} \qquad \boxed{11.9}$$

TABLE 11.2 ANOVA Table for Simple Linear Regression

Source of Variation	Degrees of Freedom	Sum of Squares	Mean Square	F
Regression	1	*SSR*	*MSR* = *SSR*/1	*MSR*/*MSE*
Error	$n - 2$	*SSE*	*MSE* = *SSE*/(n − 2)	
Total	$n - 1$	Total *SS*		

t value from Appendix table 8, we reject the null hypothesis and state that a linear relationship is described by the regression line of Y on X; otherwise not.

Analysis of Variance for the Regression Line

We partition the sums of squares and degrees of freedom associated with the random variable Y to perform the analysis of variance on the regression line. As we discussed earlier for one-way analysis of variance in Chapter 9, we get the total sum of squares as the sum of squared deviations of Y from its mean, as shown in equation 11.10. We partition it into two sums of

$$Total\ SS = \Sigma(Y_i - \overline{Y})^2 \tag{11.10}$$

squares—sum of squares for regression, *SSR*, and sum of squares error, *SSE*. We get the sum of squares for regression by summing the squared deviations of the estimated values of Y from the equation, Y_e, from the mean of Y, as in equation 11.11,

$$SSR = \Sigma(Y_e - \overline{Y})^2 \tag{11.11}$$

and compute the sum of squares error as the squared deviations around the regression line (equation 11.12). The degrees of freedom for the *Total SS* are $(n - 1)$ while *SSR* has k degrees of freedom and *SSE* has $(n - k - 1)$ degrees of

$$SSE = \Sigma(Y_i - Y_e)^2 \tag{11.12}$$

freedom. Once the sums of squares are obtained, they can be placed in the ANOVA table and the analysis completed (table 11.2).

Most computations performed on a personal computer will provide an analysis of variance table like 11.2. However, if we must use a handheld calculator, we can use some of the computational equations that follow. The total sum of squares may be computed as in equation 11.13 and the regression sum of squares may be calculated using one of the following three

$$Total\ SS = \Sigma Y_i^2 - \frac{(\Sigma Y_i)^2}{n} = \Sigma Y_i^2 - n\overline{Y}^2 \tag{11.13}$$

equations indicated in equation 11.14. The error sum of squares may be calculated by

$$SSR = b^2\Sigma(X_i - \overline{X})^2 = b[\Sigma(X_i - \overline{X})(Y_i - \overline{Y})] = b\left[\Sigma X_iY_i - \frac{\Sigma X_i \Sigma Y_i}{n}\right] \quad \boxed{11.14}$$

subtraction or by the formula in equation 11.15.

$$SSE = TotalSS - SSR = \Sigma Y_i^2 - a\Sigma Y_i - b\Sigma X_iY_i \quad \boxed{11.15}$$

We compare the computed F from the ANOVA table to critical F with 1 degree of freedom in the numerator and $(n - 2)$ in the denominator and the chosen level of significance. If the computed F is larger than the critical F from Appendix table 10, we reject the null hypothesis of no linear relationship between X and Y; otherwise not.

The Coefficient of Determination, r^2

The numerical measure of the percent of the variability in the random variable Y explained by the regression line is the coefficient of determination, written r^2. For simple linear regression, it is the square of the correlation coefficient r, which we presented earlier, but r^2 measures how well the least squares regression line fits the observed data. Thus, it is calculated as the ratio of sum of squares for regression to total sum of squares from the analysis of variance, as in equation 11.16.

$$r^2 = \frac{SSR}{Total\ SS} = \frac{b^2\Sigma(X_i - \overline{X})^2}{\Sigma Y_i^2 - n\overline{Y}^2} \quad \boxed{11.16}$$

The coefficient of determination can also be expressed as in equation 11.17. The larger the

$$r^2 = 1 - \frac{SSE}{Total\ SS} \quad \boxed{11.17}$$

value of r^2, the better the regression line fits the data. However, with certain types of data, such as prices, incomes, and socioeconomic data that agricultural economists must use, it is not uncommon to obtain coefficients of determination of 0.6 or so, which are considered quite good. The value of r^2 that is considered good depends greatly upon the type of data being analyzed. Thus, this is a common summary statistic presented with the estimated regression equation.

The Standard Error of the Estimate, $S_{y \cdot x}$

The regression equation is used primarily to estimate the random dependent variable given a certain value of X. Once we have the estimated Y value, we want to know how dependable it is. This is measured by how closely the data points are scattered around the regression line in a scatter diagram. The closer the scatter of points around the line, the more dependable the estimated Y value from the equation.

The **standard error of the estimate,** $S_{y \cdot x}$, provides a numerical measure of the dispersion of the points around the regression equation just like the scatter diagram provides a visual interpretation. The smaller the value of the standard error of the estimate, the more dependable the estimate of Y obtained from the regression equation, i.e., the closer Y_e will be to the data point Y_i for a given value of X and the less dispersion there is of the data around the regression line. Thus, in the extreme case in which all of the data points Y_i fall on the regression line, there is no dispersion, $r^2 = 1.0$ and $S_{y \cdot x} = 0$. The equation for the standard error of the estimate is given in equation 11.18,

$$S_{y \cdot x} = \sqrt{MSE} = \sqrt{(SSE)/(n - 2)} \tag{11.18}$$

where MSE is the mean square error from the ANOVA table.

The standard error of the estimate is theoretically similar to the standard deviation. Thus, its interpretation is also similar. If the scatter of Y values about the regression equation is normally distributed and we have a large sample of data, about 68 percent of the points in the scatter diagram will fall within one standard error of the estimate above and below the regression line, about 95.5 percent of the points will fall within two standard errors of the estimate above and below the line, and almost all, or 99.7 percent, of the points will fall within three standard errors of the estimate above and below the line. Thus, we can construct confidence bands around the regression line based on the standard error of the estimate, as well as confidence limits for the mean of Y.

Confidence Interval for $E(Y_e)$

In regression analysis, we may be most interested in estimating the mean response for Y given a particular value of X. We consider the mean because the data set we used to estimate the regression equation is from one of many random samples that could have been collected for values of the random variable Y given a set of X values. If we took all possible random samples and estimated regression lines from each one, the lines would not necessarily have the same intercepts a or slopes b. Thus, we could estimate many different regression lines; but all would go through the point, mean of X, mean of Y. Thus, for values of X close to its mean, the estimated values of Y obtained from each of the possible regression lines (Y_e's) would be much more similar than they would for values of X farther away from its mean. We obtain the mean of Y for a given value of X from the mean of the sampling distribution of Y_e. When we take just one sample, and compute a single regression line based on the data in that sample, we may construct a confidence interval for the mean of Y based on the t distribution, as shown in equation 11.19, in which $S(Y_e)$ is defined in

$$Y_e - t_{(\alpha/2,\, \nu)} \cdot S(Y_e) \leq E(Y_e) \leq Y_e + t_{(\alpha/2,\, \nu)} \cdot S(Y_e) \tag{11.19}$$

equation 11.20. Notice that $S(Y_e)$ as computed from equation 11.20 becomes larger for values of X_e farther away from the mean, as that difference is squared in one of the terms under the radical. This is consistent with our previous discussion.

$$S(Y_e) = S_{y \cdot x}\sqrt{\frac{1}{n} + \frac{(X_e - \overline{X})^2}{\Sigma(X_i - \overline{X})^2}} \tag{11.20}$$

An Example Problem

A farm supply store manager wishes to predict monthly sales, Y, for her company using advertising expenditures, X. She has collected 10 months of data on past performance (table 11.3). Compute the regression of Y on X, test the slope coefficient to see if it is different from zero at the 1 percent level of significance, compute r^2 and interpret it, and provide a 95 percent confidence interval estimate for the mean of Y when X_e = \$1,000.

To begin this problem, we set up the null and alternative hypotheses:

$$H_o\text{: } \beta = 0$$
$$H_a\text{: } \beta > 0$$

We use a one-sided test to the right because we expect monthly sales to increase as advertising expenditures increase. We will test these hypotheses with both the t test and analysis of variance. First, however, we compute the estimators for the regression line, a and b, with b first:

$$b = \frac{\Sigma X_i Y_i - \frac{(\Sigma X_i)(\Sigma Y_i)}{n}}{\Sigma X_i^2 - \frac{(\Sigma X_i)^2}{n}} = \frac{924.8 - \frac{(9.4)(959)}{10}}{9.28 - \frac{(9.4)^2}{10}} = 52.567$$

$$a = \overline{Y} - b\overline{X} = 95.9 - (52.567)(0.94) = 46.486$$

The regression equation estimated by the method of least squares is thus:

$$Y_e = 46.49 + 52.57X$$

To test the null hypothesis about β with ANOVA, we first calculate the total and regression *SS:*

$$\textit{Total SS} = \Sigma Y_i^2 - \frac{(\Sigma Y_i)^2}{n} = 93{,}569 - \frac{(959)^2}{10} = 1{,}600.9$$

$$SSR = b\left[\Sigma X_i Y_i - \frac{\Sigma X_i \Sigma Y_i}{n}\right] = 52.567\left[924.8 - \frac{(9.4)(959)}{10}\right] = 1{,}226.93$$

Next, we obtain the error sum of squares by subtraction:

$$SSE = \textit{Total SS} - SSR = 1{,}600.9 - 1{,}226.93 = 373.97.$$

TABLE 11.3 Advertising Expenditures and Monthly Sales (in $1,000s) for a Farm Supply Store

Advertising Expenditures, X	Monthly Sales, Y	XY	X^2	Y^2
1.2	101	121.2	1.44	10,201
0.8	92	73.6	0.64	8,464
1.0	110	110.0	1.00	12,100
1.3	120	156.0	1.69	14,400
0.7	90	56.0	0.49	8,100
0.8	82	65.6	0.64	6,724
1.0	93	93.0	1.00	8,649
0.6	75	45.0	0.36	5,625
0.9	91	81.9	0.81	8,281
1.1	105	115.5	1.21	11,025
9.4	959	924.8	9.28	93,569

We can now set up the ANOVA table and test the hypothesis (table 11.4).

The computed $F = 26.25$ is compared with critical $F = 11.26$ from Appendix table 10 for a 1 percent level of significance and 1 and 8 degrees of freedom. Since computed F is larger, we reject the null hypothesis of zero slope for the regression line and conclude that advertising expenditures explain more of the variation in monthly sales than the mean monthly sales by itself. How much of the variation in Y that is explained is given by the coefficient of determination:

$$r^2 = \frac{SSR}{Total\ SS} = \frac{1{,}226.93}{1{,}600.90} = 0.77$$

Thus, 77 percent of the variation in monthly sales is explained by the amount of monthly advertising expenditures.

TABLE 11.4 ANOVA Table for Advertising Expenditures and Monthly Sales Regression Analysis

Source of Variation	Degrees of Freedom	Sum of Squares	Mean Square	F
Regression	1	1,226.93	1,226.93	26.25
Error	8	373.97	46.75	
Total	9	1,600.90		

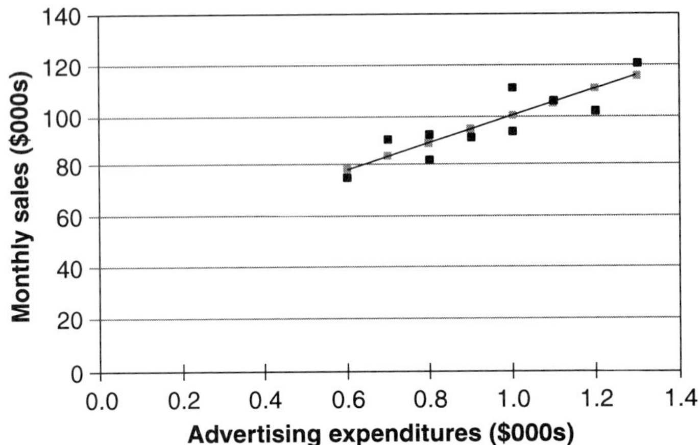

FIGURE 11.2 Estimated regression line for advertising and sales data for farm supply store

To perform a t test on β, we must first compute S_b and then calculate computed t.

$$S_b = \sqrt{\frac{MSE}{\Sigma X_i^2 - \frac{(\Sigma X_i)^2}{n}}} = \sqrt{\frac{46.75}{9.28 - \frac{(9.4)^2}{10}}} = 10.26$$

$$t = \frac{b - \beta}{S_b} = \frac{52.567 - 0}{10.26} = 5.12$$

To test the hypothesis, compare computed $t = 5.12$ with the critical $t = 2.896$ from Appendix table 8 for 8 degrees of freedom and a 1 percent level of significance. Because computed t is greater, we reject the null hypothesis of zero slope for the regression line (figure 11.2), just as before with ANOVA.

Some investigators include S_b in parentheses directly underneath b when they write the estimated regression equation. Others prefer to put computed t there and place one asterisk beside the t value if it is significant at the 5 percent level, or two asterisks if it is at the 1 percent level. If we choose either practice, we must state which value we have included so the reader will understand. For example, we could write the estimated regression equation as:

$$Y_e = 46.49 + \underset{(5.12^{**})}{52.57X}$$

and tell the audience that the number in parentheses is the computed t value.

To use the regression equation to estimate Y when $X = \$1{,}000$, we first code the X value so that it corresponds to the form used in the raw data, i.e., we let $X = 1.0$, then substitute 1.0 into the equation in place of X and solve for Y.

$$Y_e = 46.49 + 52.57X = 46.49 + 52.57(1.0) = 99$$

Thus, monthly sales are estimated by the equation to be $99,000 when advertising expenditures are $1,000. This is a point estimate. To construct a 95 percent confidence interval on the average value of sales when advertising expenditures are $1,000, we can use the confidence interval formula developed earlier from equations 11.19 and 11.20 and reproduced here.

$$Y_e - t_{(\alpha/2,\, \nu)} \cdot S(Y_e) \leq E(Y_e) \leq Y_e + t_{(\alpha/2,\, \nu)} \cdot S(Y_e)$$

where

$$S(Y_e) = S_{y \cdot x}\sqrt{\frac{1}{n} + \frac{(X_e - \bar{X})^2}{\Sigma(X_i - \bar{X})^2}}$$

To compute $S(Y_e)$, we must first calculate $S_{y \cdot x}$, which is just the square root of MSE. Thus,

$$S_{y \cdot x} = \sqrt{MSE} = \sqrt{46.747} = 6.84$$

and

$$S(Y_e) = 6.84\sqrt{\frac{1}{10} + \frac{(1.0 - 0.94)^2}{0.444}} = 2.25$$

The critical value of $t_{(0.025,\, 8)}$ from Appendix table 8 is 2.306. Plugging all of these values into the confidence interval formula gives:

$$99 - (2.306)(2.25) \leq E(Y_e) \leq 99 + (2.306)(2.25)$$
$$93.81 \leq E(Y_e) \leq 104.19.$$

When advertising expenditures are $1,000 per month, the true average monthly sales range between $93,810 and $104,190, according to our 95 percent confidence interval. This interval is wider than $10,000, which is large. In practice, if a business manager were going to do a study of this kind, she would select a larger sample than ten, which would lower the standard error of the estimate, and, thus $S(Y_e)$, which would make the confidence interval narrower.

Regression When *X* Is Random

The ordinary least squares regression model assumes that the X values are known constants. Thus, the confidence coefficients and risks of errors refer to repeated sampling when the X values are kept the same from sample to sample.

Sometimes it is not appropriate to consider the X values as fixed. For example, if we regress wheat price on wheat quantity, the quantity variable is also random because one seller cannot control the quantity sold in this market—there are too many participants, and his quantity is too small to affect the actions of those participants. Therefore, it is not meaningful to think of repeated samples where the wheat quantity sold is the same from sample to sample.

So in the case where X and Y are both random variables, are all of our earlier results not applicable? No, it can be shown that all our previous results still hold, provided that:

1. The conditional probability distributions of the Y_i given the X_i are normal and independent with means $\alpha + \beta X_i$ and variance σ^2.
2. The X_i are independent random variables whose probability distribution $f(X_i)$ does not involve the parameters α, β, σ^2, i.e., the distribution does not contain the regression parameters.

The modifications we must make because of random X variables are in the interpretation of confidence coefficients and risks of error. These now refer to repeated sampling of pairs (X_i, Y_i), where both variables change from sample to sample. Also, the power of hypothesis tests is different when X is random.

Residuals

A **residual** e_i is defined as the difference between the observed value of Y and the estimated value from the regression line (equation 11.21) and may be regarded as the observed

$$e_i \equiv Y_i - Y_e \quad \boxed{11.21}$$

error as opposed to the true error ϵ_i in the regression model (equation 11.22). For the

$$\epsilon_i = Y_i - E(Y_i). \quad \boxed{11.22}$$

regression model, we assume the ϵ_i are independent normal random variables with mean $\mu = 0$ and constant variance σ^2. If we specify the regression model properly for the data set, we expect the observed residuals e_i to have the same properties as the ϵ_i. We may use this idea as the basis for residual analysis—a useful way to examine the appropriateness of a regression model. The mean of n residuals e_i is zero, as indicated by equation 11.23. And since it is always zero, it tells

$$\bar{e} = \frac{\Sigma e_i}{n} = 0 \quad \boxed{11.23}$$

us nothing about whether the true errors ϵ_i have a mean of zero. The variance of the n residuals e_i for the regression model is equal to MSE, as shown in equation 11.24. So if the model is

$$\sigma^2 = \frac{\Sigma(e_i - \bar{e})^2}{n - 2} = \frac{\Sigma e_i^2}{n - 2} = \frac{SSE}{n - 2} = MSE \quad \boxed{11.24}$$

appropriate, the unbiased estimator of the variance of the error terms is MSE.

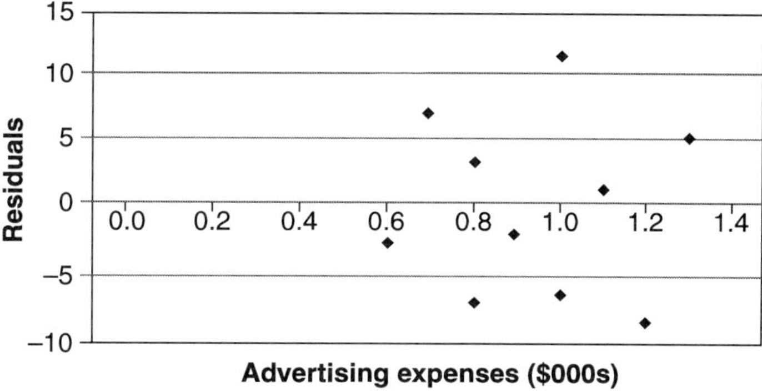

FIGURE 11.3 Scatter diagram of residuals from advertising and monthly sales regression model

If we graph the residuals in a scatter diagram against X, we can learn much about the appropriateness of the regression model from the shape of the resulting group of points. When the model is acceptable, the scatter of the points is fairly evenly distributed about zero and makes a horizontal band or glob with no apparent pattern or trend. Figure 11.3 shows this for the residuals for the example problem we did earlier for advertising and sales data.

If we should have specified a nonlinear regression function rather than a straight line, the residual plot will also form a curved line. In this case, we add a nonlinear X such as log X, or a power of X, to the regression equation as a second X variable. Not only will this increase the coefficient of determination, but the new residuals for the modified line will not indicate a curve.

Another factor we may examine with a scatter diagram of residuals is whether the error variance is constant. As the X values increase, if the scatter points in the vertical direction become farther from zero so that they make a megaphone-shaped or trapezoidal pattern, then the error variance is larger for large values of X and is not constant. We can also find cases in which the error variances decrease as X increases, which leads to a trapezoid that is larger for small values of X. Both of these conditions frequently occur with **cross section** data. If we use a multiple regression model with more than one X variable, then a residual plot against the Y_e values on the horizontal axis will also be trapezoidal when the error variance is not constant. Weighted least squares models, which are beyond our scope, will help correct for nonconstant error variance.

Whenever the data in the regression model are observations over time (a time series), it is a good idea to plot the residuals against time on the horizontal axis. If our plot is linear with first negative values of the residuals and then positive values, or if it has a definite hormonal pattern where we can draw a line through the residuals that oscillates around zero, then the error terms are correlated over time. We can use a Durbin-Watson test to indicate the amount of correlation. Error terms that are not independent cause concern, but the suggested remedies, such as using lagged variables, are beyond our scope. If the error terms are independent with time, they will have no pattern and will fall in a random horizontal group around zero.

Finally, if we are considering whether to include a particular independent variable in a regression model to increase its explanatory power, we may wish to plot the residuals against it. If the residuals exhibit systematic variation so that they are mostly all positive or negative, or form some pattern with the variable, then it might be appropriate to include that variable in the model. On the other hand, if the residuals are randomly distributed around zero and make a horizontal band, then including the variable in the model will not help much.

Multiple Regression

While we may find simple linear regression models that relate Y to one explanatory variable X to be powerful and useful, we may need to add more X variables to our model. Generally, this allows us to more fully explain the variability in Y and to improve the model's prediction and estimation. In this way, we can study three or more variables together rather than two, but the calculations are much more complex, as is interpretation of the results. We may reduce the computational effort by resorting to computer programs, but that doesn't help much with interpretation.

Multiple Regression Models

Multiple linear regression models are best depicted with matrices. However, matrix algebra is generally not a required subject for students in the agricultural sciences, so we will use an algebraic approach in the explanation that follows. Students who wish to study this topic further, however, are urged to learn matrix algebra and to take a specialized course in regression that is developed along those lines.

A **multiple linear regression** equation follows. The X variables in the equation can take on any form. Thus X_{3i} might be the product of X_{1i} and X_{2i}, or it might be the log of X_{2i} or its square, etc. We often designate models with interaction and higher order terms as complete models and name those lacking such terms reduced effects or main effects models. As long as an equation is linear in its regression coefficients (β_1, β_2, etc.), we consider it a linear equation. We use β_0 for the constant term in place of α just to make the notation more consistent (equation 11.25).

$$Y_i = \beta_0 + \beta_1 X_{1i} + \beta_2 X_{2i} + \dots + \beta_k X_{ki} + \epsilon_i \text{ for } i = 1, 2, \dots, n. \quad \boxed{11.25}$$

The assumptions for the multiple regression model are essentially the same as those for the simple linear model:

1. The error term ϵ_i has mean 0 and variance σ^2.
2. The error terms are uncorrelated.
3. $\beta_0, \beta_1, \ldots, \beta_k$ are $(k + 1)$ regression parameters, and $X_{1i}, X_{2i}, \ldots, X_{ki}$ are k known constants or random variables.
4. For purposes of testing hypotheses and estimating confidence intervals, the ϵ_i are normally distributed.

The assumptions imply that the mean of Y or the expected value, $E(Y_i)$, for a given set of X values is equal to the regression equation 11.26, which is the equation for a hyperplane in $(k + 1)$

$$E(Y_i) = \beta_0 + \beta_1 X_{1i} + \beta_2 X_{2i} + \ldots + \beta_k X_{ki} \tag{11.26}$$

dimensional space since Y is included in one dimension along with the k X's. To make this simpler, we can reduce our linear model to three-dimensional space by working with only two X's. In this case, the model describes a plane surface in that space (figure 11.4). We derive the normal equations by least squares as in the section on simple linear regression, except that there are more—an additional one for each X variable added to the model. Thus, in the case of two X's the normal equations are specified as in equation 11.27.

$$\begin{aligned} \Sigma Y_i &= nb_0 + b_1 \Sigma X_{1i} + b_2 \Sigma X_{2i} \\ \Sigma X_{1i} Y_i &= b_0 \Sigma X_{1i} + b_1 \Sigma X_{1i}^2 + b_2 \Sigma X_{1i} X_{2i} \\ \Sigma X_{2i} Y_i &= b_0 \Sigma X_{2i} + b_1 \Sigma X_{1i} X_{2i} + b_2 \Sigma X_{2i}^2 \end{aligned} \tag{11.27}$$

They can be solved simultaneously for the least squares estimators, b_0, b_1, and b_2 of the β_i since we have three equations in three unknowns. To illustrate the method of least squares in multiple regression, we consider the following example.

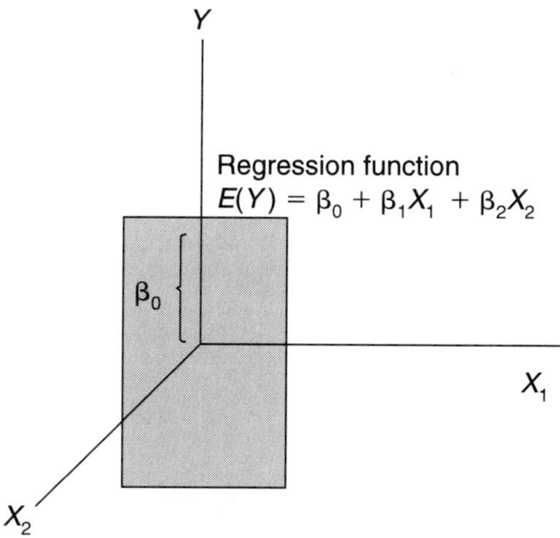

FIGURE 11.4 Regression plane for two X variables

TABLE 11.5 Multiple Regression Data for Cotton Yield, Irrigation Amount, and Fertilizer Level

Cotton Yield, Y	Irrigation Amount, X_1	Fertilizer Level, X_2
420	5	30
458	9	30
400	4	25
510	9	40
550	10	40
480	7	35
575	12	50
460	6	35
530	9	45
610	12	55
590	12	55
524	9	45

An Example Problem

An agronomist wishes to estimate the relationship between cotton yield in pounds of lint per acre (Y) and the amount of irrigation water applied in acre-inches (X_1) along with the pounds of fertilizer applied (X_2) for an irrigated farming area. She obtains a random sample of twelve farms, which provides the data in table 11.5. Estimate the linear regression model, test the regression plane and the partial regression coefficients to see if they are different from zero, and provide an estimate of the multiple coefficient of determination. Also, interpret the results.

To solve this problem, we input the data into a computer program such as Excel. We see first that the regression model is presented in equation 11.28. The computed t values for

$$Y_e = 248.063 + 10.312X_{1i} + 4.285X_{2i} \qquad R^2 = 0.97 \qquad \textbf{11.28}$$
$$(3.88^{**}) \qquad (6.27^{**})$$

both partial regression coefficients are shown in parentheses under each coefficient in equation 11.28. They are different from zero at the 1 percent level of significance since their P values are less than 0.01, thus the double asterisks. Ninety-five percent confidence intervals on the regression coefficients are:

$$4.304 \leq \beta_1 \leq 16.319$$
$$2.740 \leq \beta_2 \leq 5.831$$

The standard error of the estimate, which we use to determine these confidence intervals, is computed the same way as that for simple linear regression, but the notation is different (equation 11.29). An analysis of variance performed on the regression model tells us that the regression

$$S_{y \cdot x_1, x_2} = \sqrt{MSE} \quad \boxed{11.29}$$

plane as a whole has a slope different from zero since the computed $F = 164.389$ is significant at the 0.01 level of probability (see the appendix to this chapter for the computer output for this model).

The coefficient of multiple determination of $R^2 = 0.97$ signifies that the linear model explains 97 percent of the variability in cotton yield. We calculate R^2 for multiple regression in the same manner as that for simple linear regression (restated in equation 11.30).

$$R^2 = SSR/Total\ SS. \quad \boxed{11.30}$$

The plots of the residuals are also part of the computer output. Residuals plotted against both X variables seem to be in a normal box pattern. As we discussed in the previous chapter, residual plots are an important way of validating the linear model.

The interpretation of the partial regression coefficients is, for example for $b_1 = 10.312$, as follows. For every increase in the amount of irrigation water applied by 1 acre-inch, cotton yield increases by 10.312 pounds of lint per acre, holding fertilizer amount (X_{2i}) constant. If the b coefficient were negative rather than positive in sign, we could increase X_{1i} by one unit and cause a decrease in Y_e by the amount of b_1, holding the other X variable constant.

Multicollinearity

The preceding interpretation of the partial regression coefficients is not appropriate when the X variables are highly correlated among themselves. When the independent variables in the linear model are highly intercorrelated, we say that **multicollinearity** exists. In the case of multicollinearity, the least squares estimators $b_0, b_1, \ldots, b_k$ of the population regression parameters $\beta_0, \beta_1, \ldots, \beta_k$ are inefficient because the variances of their sampling distributions are much too large. Thus, the b_i provide nonsense answers with respect to both the values themselves and their signs, i.e., the knowledge we have about the linear model based on theory from our subject matter discipline, which we use to formulate and interpret the model, indicates that the estimated b values or their signs are out of line. If the multicollinearity is severe, then one of the X's can be written as a linear combination of a subset of the others and the regression model really contains two estimates of the same data. The solution in this instance is to delete the offending X variable from the model, but for most cases, the linear relationship among the X's is not that pronounced and more advanced techniques that are beyond our scope must be used. For more information on this topic, refer to an advanced text on multiple regression or econometrics.

Time Series

Researchers, as well as managers and marketers of agricultural products, are interested in time series because they help explain what has happened in the past and are useful in forecasting future events. Time series analysis is the orderly examination of data collected over time, such as monthly prices paid by farmers for inputs since 1903. In general, a **time series** is a set of statistical data that is collected, recorded, or observed over successive periods of time. Changes in time series data occur because of the many forces acting upon them. For example, in the early 1900s, farmers did not purchase many inputs except items such as draft animal supplies and machinery. But as time progressed, commercial fertilizer became available, along with hybrid seed, tractors, and their associated equipment. Farmers found themselves purchasing more and more of the items used in their farming operations. Thus, prices paid for inputs became more and more important as the list of inputs grew. There was an incentive to track the data and use it as a measure of farmer well-being, along, with other time series on prices received for farm products. Today, we find many important economic time series, some of which are kept at the local level by individual businesses, and others that are reported by industry and national sources.

If we can assume that there are regular and recurring components that interact in predictable ways to produce a certain time series, we can analyze them for the purpose of predicting the series for some future time period. Time series analysis is a method that attempts to separate the data into its component parts of:

1. Secular trend,
2. Cyclical fluctuations,
3. Seasonal variations, and
4. Irregular movements.

Secular trend refers to the smooth upward or downward movement that characterizes a time series over a long period of time, such as 10 years or more. We attribute this to the underlying influence of the basic forces affecting the data such as population growth, technological change, and large-scale shifts in consumer tastes and preferences. Thus, trend represents a stable movement and its description requires a different type of model than those used for the other components.

Cyclical fluctuations are movements in the business cycle that reflect the expansion of the economy from a low recessionary period through an economic boom, and then its contraction toward another recession. The business cycle is fairly long, taking from 2 to 15 years to go from trough to trough, and is irregular in its expansion and contraction. It does this around a trend line. The periods of expansion and contraction depend upon the overall health of the economy and the impact of the business activity on the particular data series we are studying. Because of the nature of the business cycle, we have not devised a highly accurate method of forecasting cyclical fluctuations in time series data.

We employ a residual method, i.e., after we take trend and seasonal variation out of time series data, cyclical and irregular variations remain. And we generally do not try to separate these two. Because the cyclical movements are frequently large relative to irregular variations, this approach works.

Seasonal variations are short-term cycles within the data that complete their duration within the year and recur annually. Weather and customs may produce seasonal variations. In agriculture, for example, the harvest period for major crops stimulates buying activities by farmers that tend to recur each year at about the same time. This makes certain product sales higher at particular times every year—a seasonal pattern. Monthly or quarterly data series are usually needed to examine seasonal variations.

Irregular movements in time series data follow no particular pattern. They result from the influence of wars; weather hazards such as hail, floods, earthquakes, and tornadoes; strikes, accidents; and so on. Thus, we have no way to predict such events and we make no effort to model them. We place irregular movements in the residuals after we have removed the other factors from the data.

Secular Trend Analysis

Because time series data span a large number of years, there may have been important changes during the period due to changing population, changing price levels, or changing definitions of important terms, such as the definition of a farm for agricultural data. We may want to divide the series by the population to obtain per capita data and analyze the trend in these per capita figures if changes in population affect the trend. Or we may wish to divide money data by an index that accounts for changes in the purchasing power of the dollar over time before we estimate a trend. If the data require that a consistent definition of a farm be used, we may have to adjust the data for changes in the definition since the U.S. Department of Agriculture makes changes periodically. After we are satisfied that there are no systematic problems in the data, then we may proceed to estimate the trend.

Linear Trend

We must first decide whether the trend is linear. As an aid, we can plot the data with time on the x axis in a scatter diagram and observe the pattern that the dots make. If they fall approximately in a straight line, then we can fit a linear trend to the data using regression analysis. When we use regression, we usually code the time variable X and use the coded values to estimate the equation rather than the year values. A common approach is to code the initial year in the time series as 0, the second year as 1, and so on, until we reach the end of the data series. We then use the appropriate coded X in the regression equation to predict Y. An alternate approach is to code the middle year of the data set 0, the succeeding years as 1,2, 3, etc., and previous years as $-1,-2$, -3 and so on until we reach the beginning of the data set. With this procedure, the X values add to zero. Or if we have an even number of years, we can code the two middle observations as 1, −1 and have numerical progressions forward and backward in time as 1, 3, 5, etc. and −1, −3, −5 and so on until we reach the end of the data. The X values coded in this manner also add to zero. We use whatever coding scheme is most convenient and let the audience know which scheme we use.

Nonlinear Trends

If the time series is not linear, we use multiple regression analysis. We must not only find an equation that fits the data, but also a form that we can justify according to the underlying economic nature of the time series.

One form is the exponential curve, which describes a time series that increases or decreases by a constant proportion over time such as population growth, new product sales, or the spread of a communicable disease (equation 11.31). Its shape depends upon the values of a

$$Y = ab^X \qquad \boxed{11.31}$$

and b. The value of a is the Y intercept, but the curve decreases for b between 0 and 1, and it increases for b values greater than 1. We estimate the equation by taking the logarithms of both sides, as in equation 11.32. This equation is linear in the logarithms and we can estimate it by

$$\log Y = \log (ab^X) = \log a + X \log b \qquad \boxed{11.32}$$

the method of least squares. To estimate the equation, we code X as we discussed in the previous section, say from 0 to n for our n years of data, and then take the logarithms of Y for each year. We run the regression on coded X and log Y. The estimated values of the regression coefficients are thus log a and log b. We get an estimated Y by placing the appropriate coded X in the equation to get log Y and then we take the antilog (log Y) to obtain the estimated Y value.

Consider the following example. We obtain annual sales data for fertilizer in a farming region (table 11.6) and wish to estimate the trend by fitting an exponential curve. We first code the X values and take the logarithms of the Y values, then estimate the exponential equation using linear regression. The estimated equation is presented in equation 11.33, with the number in parentheses being the computed t value, which is significant at the 0.01 level.

$$\underset{\phantom{(334.6^{**})}}{\text{Log } Y = 1.4153 + \underset{(334.6^{**})}{0.0334X}} \qquad \boxed{11.33}$$

To estimate the trend value for 1997, we place the coded value of X for 1997 into the preceding equation and solve for Log Y and then take the antilog of the result:

$$\text{Log } Y = 1.4153 + 0.0334(8) = 1.6825$$
$$\text{Antilog } (1.6825) = 48.1.$$

TABLE 11.6 Time Series Data for Fertilizer Sales, 1989–1996, in Raw and Coded Form

Time, *X*	Fertilizer Sales, *Y*	Coded *X*	Log *Y*
1989	26.0	0	1.4150
1990	28.1	1	1.4487
1991	30.3	2	1.4814
1992	32.8	3	1.5159
1993	35.4	4	1.5490
1994	38.3	5	1.5832
1995	41.2	6	1.6149
1996	44.5	7	1.6484

Thus, we expect fertilizer sales to be $48.1 million dollars for this region in 1997 if they follow the trend. We also plotted the regression line to the data in figure 11.5. First, notice that the data points are approximately linear once we transformed them to logarithms for Y and coded values for X, so it is no surprise that the regression line fits so well.

In addition to an exponential curve, we can fit a second-degree polynomial, a form of the quadratic equation, to nonlinear trend data (equation 11.34), which is generally appropriate when

$$Y = b_0 + b_1X + b_2X^2 \tag{11.34}$$

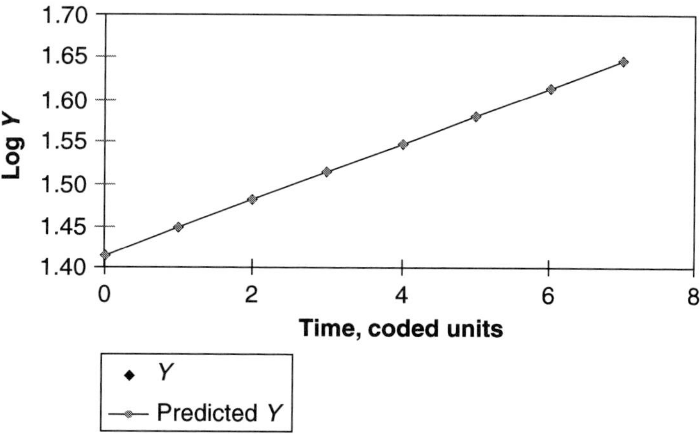

FIGURE 11.5 Line fit for nonlinear trend

there is a constant rate of change in the value of Y over time. To help determine the rate of change in Y, first compute the change in Y from one year to the next by subtraction. Next, for the change data just computed, subtract one year from the next. If the rate of change, which is the result of the second subtraction, is almost the same for the entire period, then a second-degree polynomial will fit the data well.

Other nonlinear forms that are often used to fit trends include the growth, or Gompertz, curve, the logistic curve, and the modified exponential curve, but we will not explore these.

Seasonal Variation

The most common method of smoothing data or estimating seasonal variation is the ratio-to-moving average method. This method requires that we have monthly or quarterly data in a time series. We describe the steps in this procedure for quarterly data; however, they are logically the same for monthly data. The steps are:

1. Compute a four-quarter moving average that contains the trend and cyclical factors present in the original data. The moving average is an annual average of the original quarterly data advanced one quarter at a time. That is, we average the first four quarters in the time series, drop the first quarter, add the fifth quarter, and average those four quarters, drop the second quarter, add the sixth and average those four, etc., until we have finished with the data series.
2. Next, divide the original data for each quarter by its corresponding moving average to obtain a "ratio-to-moving average" number that now contains only seasonal and irregular components, since the moving average eliminated trend and cyclical factors.
3. Rearrange the "ratio-to-moving average" numbers by quarters so that all Quarter I numbers are together, all Quarter II values are together, etc. Average these values in an attempt to eliminate the irregular component in the data. We use a modified mean for this purpose, i.e., we drop the highest and lowest values and average the remaining ones. We obtain four modified means—one for Quarter I, another for Quarter II, etc.
4. We adjust the modified means so that they add to 400 and, thus, average out to be 100 each. The adjusted modified means constitute the "seasonal index," which we use to seasonally adjust the original data and, thus, remove the influence of seasonal factors from the time series.

Consider the following example. Quarterly exports of vegetables and vegetable products from 1992 to 1996 are given in table 11.7 in billions of dollars. Compute a seasonal index for these data.

In the first step of computing a seasonal index, we obtain the four-quarter moving total for the time series. So for the 1992 data, we add 1.86, 2.20, 2.42, and 1.91 to get our total of 8.39 and place it near the center of the data, across from Quarter III. To get the next moving total, 8.45, which goes opposite Quarter IV, we drop off the first value, 1.86, and include the next value in the series, 1.92. We continue in this manner until we have computed all of the moving totals.

In the next step, we compute the four-quarter moving average by dividing each moving total value by 4 and recording the answer in the adjacent column of the table.

In the third step, we compute the specific seasonals by dividing the original data by the four-quarter moving average and multiplying by 100 percent so the result is expressed as a percent.

Now we are ready to compute the seasonal index. To do this, we write the specific seasonals in table 11.7 by year and quarter. We now have four values for each quarter. Since we want only one, we average the four to get a **modified mean** by striking out the lowest and highest values in each column and

TABLE 11.7 Quarterly Exports of Vegetables and Vegetable Products, Billion Dollars

Year and Quarter	Vegetable Exports	Moving Four-Quarter Total	Moving Average	Ratio of Original Data to Moving Average, Percent
1992 I	1.86			
II	2.20			
III	2.42	8.39	2.10	115.2
IV	1.91	8.45	2.11	90.5
1993 I	1.92	8.59	2.15	89.3
II	2.34	8.68	2.17	107.8
III	2.51	8.75	2.19	114.6
IV	1.98	8.66	2.16	91.7
1994 I	1.83	8.47	2.12	86.3
II	2.15	8.06	2.02	106.4
III	2.10	7.88	1.97	106.6
IV	1.80	7.89	1.97	91.4
1995 I	1.84	7.76	1.94	94.8
II	2.02	7.96	1.99	101.5
III	2.30	8.12	2.03	113.3
IV	1.96	8.35	2.09	93.8
1996 I	2.07	8.74	2.18	95.0
II	2.41	8.78	2.20	109.5
III	2.34	8.79	2.20	106.4
IV	1.97			

TABLE 11.8 Ratios of Original Data to Moving Average by Quarter

Year	Quarter I	Quarter II	Quarter III	Quarter IV
1992			~~115.2~~	~~90.5~~
1993	89.3	107.8	114.6	91.7
1994	~~86.3~~	106.4	~~106.6~~	91.4
1995	94.8	~~101.5~~	113.3	~~93.8~~
1996	~~95.0~~	~~109.5~~		
Mean	92.0	107.1	114.0	91.6

TABLE 11.9 Calculation of a Seasonal Index

Quarter	Modified Mean	Seasonal Index
I	92.0	90.9
II	107.1	105.9
III	114.0	112.7
IV	91.6	90.5

averaging the remaining values. These means are almost our seasonal index, but not quite—they don't add to 400. In this example, the total of the four modified means is 404.7. We force them to add to 400 by taking the ratio 400/404.7 = 0.9884 and multiplying each mean by this correction factor to get the seasonal indexes (table 11.7) Now that we have the seasonal indexes as listed in column three of table 11.7, we take these and deseasonalize our original vegetable export data (table 11.8). We take the original data and divide each value by the seasonal index divided by 100 for the appropriate quarter. For example, for Quarter I of 1992, the amount of exports is $1.86 billion. The seasonal index for Quarter I is 90.9 (table 11.9). So we divide 90.9 by 100 to get 0.909, and divide 1.86 by that to get the deseasonalized value of $2.05 billion (table 11.10). We repeat the process until we finish the time series.

Endnote

[1]This statement is a direct result of the *Gauss-Markov theorem*, which says that under conditions of the model $Y_i = a + bX_i + \epsilon_i$ where ϵ_i is the deviation $(Y_i - Y_e)$, the least squares estimators a and b are unbiased and have minimum variance among all unbiased linear estimators.

TABLE 11.10 **Deseasonalized Exports of Vegetables and Vegetable Products, Billion Dollars**

Year and Quarter		Vegetable Exports	Seasonal Index	Deseasonalized Exports
1992	I	1.86	90.9	2.05
	II	2.20	105.9	2.08
	III	2.42	112.7	2.15
	IV	1.91	90.5	2.11
1993	I	1.92	90.9	2.11
	II	2.34	105.9	2.21
	III	2.51	112.7	2.23
	IV	1.98	90.5	2.19
1994	I	1.83	90.9	2.01
	II	2.15	105.9	2.03
	III	2.10	112.7	1.86
	IV	1.80	90.5	1.99
1995	I	1.84	90.9	2.02
	II	2.02	105.9	1.91
	III	2.30	112.7	2.04
	IV	1.96	90.5	2.17
1996	I	2.07	90.9	2.28
	II	2.41	105.9	2.28
	III	2.34	112.7	2.08
	IV	1.97	90.5	2.18

Appendix to Chapter 11: Correlation and Regression with Excel

The Excel program has separate procedures for correlation and for regression. We first work the example problem we used with correlation using Excel and then two regression examples.

Correlation

Let's examine the beef carcass quality grade scores, X, and selling price per pound, Y, with the Correlation procedure using the Data Analysis submenu of the Tools Menu in Excel. First, input the data into separate columns in the spreadsheet with

TABLE 11.11 Correlation Output from the Excel Spreadsheet

	Quality Grade Score	Selling Price per Pound
Quality grade score	1	
Selling price per pound	0.93873986	1

data labels in the first row, then select the procedure. On the pop-up table, input the range of the data in the first box, check the data labels box, indicate the output range, and click OK. The resulting output is presented in table 11.11.

Notice that the correlation coefficient is the same correlation coefficient that we obtained earlier except for rounding (0.9387 or 0.94). There is no provision in the program to test the coefficient.

Simple Regression

We can use the farm supply store example problem to look at the simple regression procedure with Excel. First, input the data in the spreadsheet and then select the Regression procedure from the Data Analysis submenu of the Tools

TABLE 11.12 Simple Regression Output from the Excel Spreadsheet

Summary Output

Regression Statistics	
Multiple *R*	0.875442
R Square	0.766398
Adjusted *R* Square	0.737198
Standard Error	6.83715
Observations	10

ANOVA

	df	*SS*	*MS*	*F*	Significance *F*
Regression	1	1226.927	1226.927	26.24633	0.000904
Residual	8	373.973	46.74662		
Total	9	1600.9			

	Coefficients	Standard Error	*t* Stat	*P* Value	Lower 95%	Upper 95%
Intercept	46.48649	9.884566	4.702936	0.001536	23.69262	69.28035
X Variable 1	52.56757	10.26086	5.123117	0.000904	28.90597	76.22916

Menu. In the pop-up table, input the Y data range in the first box and the X data range in the second box, check the labels box and the 95 percent confidence level box, input the output range, check the residuals box and the residual plot box and perhaps the line plot box, and then click OK. The output is in table 11.12. Notice that it is the same as the output we obtained earlier using our equations and a calculator except for rounding.

A Multiple Regression Problem

The example problem in the section on multiple regression with data presented in table 11.6 can be solved easily with Excel. The procedure is the same as that for simple regression and the computer output is provided in table 11.13. Notice that the analysis of variance table contains 2 degrees of freedom for regression since there are two X variables in the model. Otherwise the output is similar in format to that presented for the simple regression model. The residuals were plotted by the computer and are presented in the following two graphs, one for each X variable.

They indicate no observable problems with the data since they each form a horizontal band around zero.

TABLE 11.13 Computer Multiple Regression Output for Cotton Yield Linear Model

Summary Output

Regression Statistics	
Multiple *R*	0.986588
R Square	0.973355
Adjusted *R* Square	0.967434
Standard Error	12.05794
Observations	12

ANOVA

	df	*SS*	*MS*	*F*	Significance *F*
Regression	2	47,802.37	23,901.19	164.3892	8.23*E*-08
Residual	9	1,308.545	145.3939		
Total	11	49,110.92			

	Coefficients	Standard Error	*t* Stat	*P* Value	Lower 95%	Upper 95%
Intercept	248.063	15.29534	16.2182	5.72*E*-08	213.4625	282.6635
X Variable 1	10.31164	2.655555	3.883045	0.003714	4.304352	16.31893
X Variable 2	4.285477	0.683127	6.273323	0.000146	2.740135	5.83082

TABLE 11.13 (Continued)

Residual Output

Observation	Predicted Y	Residuals
1	428.1855	−8.18553
2	469.4321	−11.4321
3	396.4465	3.553494
4	512.2869	−2.28687
5	522.5985	−7.40149
6	470.2362	9.763797
7	586.0766	−11.0766
8	459.9246	0.075438
9	533.7143	−3.71426
10	607.504	2.496044
11	586.8807	3.119326
12	533.7143	−9.71426

Exercises

1. What is the difference between regression and correlation analysis in the selection of the X values? Can we take a correlation problem and make it a regression, or vice versa? Discuss.
2. What is the "standard error of the estimate"? Is it a parameter or a statistic? Discuss.
3. What do the points on a scatter diagram represent? What does a point on a sample least squares regression line mean? Discuss.
4. The following data show national net farm income, X, and total non-real estate assets on farms, Y, in billions of dollars for the years 1988 through 1996. Calculate the correlation coefficient r and test at the 5 percent level to see if it is different from zero. Interpret the coefficient in the context of this problem.

Year	X Net Farm Income (in billions)	Y Total Non-real Estate Assets (in billions)
1988	$38	$204
1989	$45	$212
1990	$44	$220
1991	$38	$215
1992	$47	$226
1993	$43	$231
1994	$48	$237
1995	$36	$221
1996	$52	$227

5. The planted acres of rice, X (in millions), is used to predict total production, Y (in million cwt.), in regression analysis. Compute the least squares regression line for these data.

X, Acres Planted (in millions)	Y, Production (in million cwt.)
3.2	180
2.9	156
3.4	198
3.1	174
2.8	171
3.1	180

a. Test to see if the slope coefficient is different from zero using analysis of variance and a 5 percent level of significance. Interpret the slope coefficient in the context of this problem.
b. Compute the coefficient of determination and interpret it in the context of this problem.
c. Predict production when acreage planted is 3.0 million acres using the equation for the regression line. Construct a 95 percent confidence interval on the predicted production mean. Interpret the confidence interval in the context of this problem.
d. Plot the scatter diagram for these data and draw in the least squares regression line.

6. The following data relate per capita personal income, X (in thousands of 1992 dollars), to per capita beef consumption, Y (in pounds).

X Income (in thousands)	Y Beef Consumption (in pounds)
17.6	69
17.8	65
17.9	64
17.8	63
18.1	62
18.1	61

a. Compute the least squares regression line of Y on X.
b. Use a t test to determine if the slope of the line is different from zero at the 0.05 level of significance.
c. Estimate beef consumption when per capita income is 18.0 thousand dollars. Place a 95 percent confidence interval on this estimate and interpret it in the context of this problem.
d. Calculate r^2 and interpret it in the context of this problem.
e. Plot a scatter diagram for these data and draw in the least squares regression line.

7. Use Excel to work exercise 4.

8. Use Excel to work exercise 5.

9. Use Excel to work exercise 6.

10. An agronomist obtains experimental data on the yield per acre of cotton, Y (in pounds of lint), and amount of nitrogen fertilizer, X_1 (in pounds), and amount of irrigation water applied, X_2 (in acre-inches). Estimate the least squares regression line using Excel.

Y Cotton Yield (in pounds)	X_1 Nitrogen (in pounds)	X_2 Irrigation (in acre-inches)
250	0	0
320	10	0
375	15	3
400	20	3
410	25	3
440	25	6
450	30	6
470	35	6
510	35	9
530	40	9
532	45	9

a. State the regression equation and investigate whether the overall model is useful. Test using $\alpha = 0.05$.

b. Test the significance of the slope coefficients using a 5 percent level of significance.

c. State R^2 and interpret it in the context of this problem. State the standard error of the estimate.

11. A small agricultural manufacturing firm wants to estimate its weekly total cost function. The manager obtains accounting data for the firm and gleans the following relationship. Estimate the equation for total costs using a multiple regression equation that contains X and X^2 and answer parts a. through c. under exercise 10 for this problem.

Y Total Costs	X Weekly Output
$4,800	0
$4,815	5
$4,830	10
$4,850	15
$4,875	20
$4,900	25
$4,925	30
$4,950	35
$5,000	40
$5,060	45

12. The sales of corn seed are related to the number of days of radio and TV advertising on shows catering to farmers and the amount spent on the advertising campaign as follows. Use Excel to determine the regression equation.

a. Test the slope coefficients to see if they are different from zero at the 5 percent level of significance.
b. Determine the multiple coefficient of determination and interpret it in the context of this problem.
c. Predict sales if the advertising campaign runs 10 days and costs $24,000.
d. What might be the problem with these data? What, if anything, provides such a clue?

Y Sales of Corn Seed	X_1 Number of Days	X_2 Amount Spent
$000s		$000s
$235	8	$16
$245	10	$20
$260	12	$25
$270	14	$30
$255	11	$22
$265	13	$26
$275	15	$32
$240	9	$18
$233	7	$15

13. Calves in the 400–600 pound weight class (X_3) were placed on bermuda grass pasture during spring, fall, and summer ($X_2 = 1, 2,$ or 3, respectively), supplemented with grain according to body weight (X_1 = percent body weight), and pounds of gain per day (Y) was measured. Calculate the relationship between gain and these other variables using a multiple regression model.
   a. Test the slope coefficients to see if they are different from zero at the 5 percent level of significance and interpret them in the context of this problem.
   b. Compute R^2 and interpret it in the context of this problem.
   c. Predict the gain in pounds per day of a 510-pound calf fed a supplement equal to 1 percent of its body weight and grazed on summer pasture.

Y Gain, pounds/day	X_1 Season	X_2 Supplement	X_3 Body Weight
2.5	1	1.0	590
2.0	1	1.0	415
2.1	1	0.75	530
1.0	3	1.0	570
0.8	3	0.75	520
0.5	3	0.5	425
1.5	2	0.75	540
1.8	2	1.0	405
2.0	2	1.0	562

14. Use the method of least squares to estimate the linear trend line for the Farm Products Price Index (1982 = 100) from 1980 to 1995.

Year	Farm Products Index
1980	102.9
1981	105.2
1982	100.0
1983	102.4
1984	105.5
1985	106.1
1986	105.2
1987	108.3
1988	109.9
1989	108.9
1990	112.2
1991	111.7
1992	114.6
1993	113.1
1994	116.3
1995	117.4

15. Use linear regression with Excel to estimate the trend line for the data in exercise 14. Compare the two estimates. What can you say about the trend? Using the regression line, estimate the Farm Products Price Index for 1997.

16. Outstanding mortgage debt on farm property in billions of dollars from 1980 to 1993 is given in the following table. Fit a trend line to the data using an exponential curve and a straight line. Which fit the data best? Predict outstanding mortgage debt for 1995.

Year	Outstanding Mortgage Debt (in billions)
1980	$30.5
1981	$32.4
1982	$35.4
1983	$39.8
1984	$44.9
1985	$49.9
1986	$55.4
1987	$63.9
1988	$72.8
1989	$86.8
1990	$97.5
1991	$107.2
1992	$118.3
1993	$133.7

17. The following trend equation resulted from fitting a parabola by regression analysis to the size of the farm labor force in a particular county.

    $Y = 49.17 + 4.23X - 0.19X^2$
    $X = 0$ in 1975
    X is in 2½-year intervals
    Y is the size of the farm labor force in thousands

    a. If we assume the trend line fits the data well, what generalizations can be made about the way the farm labor force has grown in this county in percentage terms? In absolute amounts?
    b. In part a., we assumed that the line fits the data well. However, the actual size of the farm labor force in 2000 was 92,073. This number indicates the line is not a good fit. Do you agree? Discuss.

18. One of your managers has forecasted sales for the company next year at $12 million by using a trend line. We expect no sharp cyclical fluctuations next year and we think that the effect of trend within the year will be negligible, and we expect the seasonal pattern in sales to continue. The pattern is as follows:

Quarter	I	II	III	IV
Seasonal index	90	75	130	105

    Prepare a sales forecast for the second and third quarters of next year.

19. The index for seasonal variation in beef production for selected months is 60 for January, 70 for March, 122 for August, and 100 for September.
    a. In which of these months is beef production the greatest?
    b. In which of these months is production most typical of the monthly average?
    c. Beef production in the Southwest increased from 6.06 million pounds in January to 11.59 million pounds in August. What was the percentage change in production after making allowance for seasonal variation?

20. Commercial slaughter of young broilers exhibited the following monthly pattern (in million head).
    a. Plot these data over time.
    b. Calculate the linear trend and a monthly seasonal index using a 12-month moving average centered on the sixth month.

c. Adjust the monthly slaughter for seasonal variation.

Month	Year			
	1997	1998	1999	2000
January	33	36	40	42
February	30	31	37	36
March	35	37	37	38
April	32	31	38	41
May	33	39	39	42
June	35	41	35	39
July	34	36	41	42
August	38	38	39	41
September	37	36	36	41
October	35	38	41	45
November	34	35	39	37
December	32	34	37	42

CHAPTER

12

Nonparametric Statistics

Most of the statistical tests we have considered in previous chapters have been based on a particular probability distribution or sampling distribution such as the normal, t, F, etc., and we have made assumptions about the form of the probability distribution underlying the variables we were testing. For some problems, we are unwilling to make such assumptions about the population parameters, and, thus, we use tests that are distribution-free, or **nonparametric**.

In many such problems, the data are not quantitative in nature, but are **nominal** or **ordinal**. With nominal data, objects are named (hence, nominal), but not much else. Thus, we can have livestock sorted into calves, heifers, steers, cows, and bulls; vegetables grouped by type such as asparagus, peppers, carrots, and so on; or voters arranged by political party, etc. The form of these data is generally the number or frequencies that fall into each class. Ordinal data express a relationship of order so that ranks may be used. Thus, consumers may rank their favorite foods, the press may rank college football teams nationally, or we may see other subjective rankings of items such as personal character traits, beauty, or flavor. These types of data do not lend themselves to actual measurement because it is not possible, for example, to say how much better a consumer likes her favorite food than the food in second or third place.

Thus, when we deal with nominal or ordinal data, we use tests like those just described.

The Mann-Whitney Test

We use the Mann-Whitney test to determine whether two random samples are drawn from the same or different populations. It is the nonparametric counterpart of the two independent sample t tests. But it does not require that we assume the two population distributions are normal as the t test does. The Mann-Whitney test is based on the concept that if the two independent random samples are drawn from the same population, then the average of the sample ranks $r(X_{1i})$ and $r(X_{2i})$should be equal. On the other hand, if the average of the ranks $r(X_{1i})$ from the first sample, is not the same as the average of the ranks $r(X_{2i})$ from the second sample, then we most likely have samples from two different populations, and we reject the null hypothesis of equal sample means.

To do the test, we take the data from each sample and rank it as though all of it comes from a single pooled sample, and we treat ties in the data by averaging the ranks. We sum the ranks of the data for the first sample and denote their total as the statistic U. U is approximately normally distributed with mean $E(U) = [n_1 n_2]/2$ and variance $\sigma_U^2 = [n_1 n_2(n_1 + n_2 + 1)]/12$. Thus, we compute Z using equation 12.1 and reject the null hypothesis if computed Z exceeds Z_α from Appendix

$$Z = \frac{U - \frac{n_1 n_2}{2}}{\sqrt{\frac{n_1 n_2(n_1 + n_2 + 1)}{12}}} \tag{12.1}$$

table 7, as in any Z test. The computed Z is negative if U is small, which happens when the data in the first sample are less than those in the second because it is the sum of all the low ranks. Conversely, computed Z is positive when U represents the sum of all of the high ranks, which happens when the data in the first sample are larger. This test works for one-sided or two-sided alternative hypotheses.

For an example, we consider the following. Two workers are checking the percent moisture in truckloads of grain sorghum brought to the local elevator. The first worker takes ten samples during the morning, while the second gets eleven samples (table 12.1). The percent moisture readings from each worker are shown in the table. Is there a difference in their average moisture readings at the 5 percent level of significance?

TABLE 12.1 Moisture Readings for Two Workers

Worker 1		Worker 2	
14	8	6	11
12	5	7	10
15	4	8	9
13	6	6	12
9	16	4	7
		5	

First, we write the null and alternative hypothesis as:

$$H_o\colon \mu_1 - \mu_2 = 0$$
$$H_a\colon \mu_1 - \mu_2 \neq 0.$$

Since this is a two-sided test, we obtain the critical Z value from Appendix table 7 as ±1.96 for a 5 percent level of significance. Next, we rank the pooled data and identify that from the first sample by placing it in bold italics (table 12.2).

$$U = \Sigma r(X_{1i}) = 1.5 + 3.5 + 6 + 10.5 + 12.5 + 16.5 + 18 + 19 + 20 + 21 = 128.5$$

$$E(U) = [n_1 n_2]/2 = [10 \cdot 11]/2 = 110/2 = 55 \text{ and}$$

$$\sigma_U{}^2 = [n_1 n_2(n_1 + n_2 + 1)]/12 = [10 \cdot 11(10 + 11 + 1)]/12 = [110(22)]/12 = 2420/12 = 201.66.$$

Thus, computed $Z = (U - E(U))/\sigma_U = (128.5 - 55)/\sqrt{201.66} = 73.5/14.2 = 5.18$, which, when compared to the table Z, falls into the critical region. We reject the null hypothesis of equality of average moisture readings by the two workers and state that the first worker has higher average readings.

The Sign Test

The sign test is an easy test to administer and is the nonparametric equivalent of the paired *t* test for two nonindependent samples. There are other tests, such as the Wilcoxon Signed Rank test, that are also appropriate in this case. In the sign test, we usually assume ordinal data so that we record whether the first observation from each pair is higher than the second with a plus sign, and is lower than the second with a minus sign. The null hypothesis for the sign test is usually that the two samples are drawn from populations with the same median, so that the probability of a plus sign or a minus sign is 0.5 for each pair of data. Thus, both the binomial distribution and the normal approximation to the binomial are appropriate sampling distributions for this test. We employ the latter when we have large samples. We do the test for large samples ($n > 20$) with computed Z defined as in equation 12.2 where R is the number of positive signs and n represents the number of pairs of data. Compare computed Z from equation 12.2 with the table Z to accept or reject the null hypothesis.

$$Z = (2R - n)/\sqrt{n} \tag{12.2}$$

TABLE 12.2 Ranking of Moisture Reading Data

Measurement	***4***	4	***5***	5	***6***	6	6	7	7	***8***	8
Rank	1.5	1.5	3.5	3.5	6	6	6	8.5	8.5	10.5	10.5
Measurement	***9***	9	10	11	***12***	12	***13***	***14***	***15***	***16***	
Rank	12.5	12.5	14	15	16.5	16.5	18	19	20	21	

TABLE 12.3 Consumer Ratings of Two New Food Products

Consumer	French Fried Tomatoes	Dried Apricot Snack	Sign
1	15	18	−
2	12	8	+
3	4	18	−
4	19	11	+
5	10	5	+
6	17	19	−
7	13	16	−
8	9	15	−
9	14	12	+
10	16	18	+
11	11	13	−
12	15	18	−
13	8	12	−
14	12	17	−
15	16	19	−
16	17	15	+
18	13	15	−
19	18	15	+
20	15	17	−
21	16	19	−
22	11	17	−

Consider the following example. Suppose we ask a panel of twenty-two people to rate two new food products on a scale from 1 to 20, with 20 being the best. The data are shown in table 12.3. Test to see if the median rankings for the two products are the same at the 5 percent level of significance.

We write the null and alternative hypotheses as:

$$H_o: p = 0.5$$
$$H_a: p \neq 0.5$$

This is a two-sided test because we do not ask whether one product is better than the other. The number of plus signs for this data set is seven. We thus use equation 12.2 to solve for Z and compare computed Z with the critical Z from Appendix table 7 of ± 1.96 to decide whether to accept or reject the null hypothesis. Given our computed $Z = -1.71$, we cannot reject the null

$$Z = (2R - n)/\sqrt{n} = (2(7) - 22)/\sqrt{22} = (14 - 22)/4.69 = -1.71$$

hypothesis and we conclude no preference of these consumers for one new food product over the other.

The Runs Test

In the previous section on the sign test, we worked with the number of plus signs, but we might also be interested in the number of runs of plus and minus signs, in which we define a run as a sequence of numbers with the same sign. Thus, if we observe the sequence

$$- - - - + + + +$$

we note that it contains two runs, the first with four minus signs and the second with four pluses. In contrast, the sequence

$$- + - - + + - + + +$$

has six runs, the first with one minus sign, the second with one plus, the third with two minuses, the fourth with two pluses, the fifth with one minus, and the sixth with three pluses. The number of runs in a sequence can vary from one (all the same sign) to n (the signs alternate). We can test the null hypothesis of randomness by expecting the number of runs to represent a random sequence of the numbers within a data set. We reject the null hypothesis if there are too few or too many runs within the sequence.

To conduct the test, determine the median of the data set and denote observations below the median with a minus sign and those above with a plus. The number of observations that lead to a minus sign, n_1, and the number that lead to a plus sign, n_2, are equal for these data since we have calculated the median. Now count the number of runs, r, in the data and compare r to the table value in Appendix table 13 to determine whether to reject the null hypothesis.

We can use the runs test to see whether the error terms in a regression model are random. An example of this application is given in the problems at the end of this chapter.

Consider this example problem. The following sequence is supposed to be a random set of numbers with values less than 50. Use the runs test to see whether the data are random at the 5 percent level of significance. The data are:

15	6	29	48	32	12	17	39	31	4
7	16	45	49	13	20	2	25	28	33

To solve the problem, we compute the median and write down the plus and minus signs to signify values above and below the median and then determine the number of runs, r. The median is 22.5 and our list of pluses and minuses is

$$- - + + + - - + + - - - + + - - - + + +$$

The total number of runs from this test is $r = 8$. From Appendix table 13, with $n_1 = n_2 = 10$, the critical number of runs is six or less. Thus, we cannot reject the null hypothesis of randomness at the 5 percent level of significance.

Spearman's Rank Correlation

The correlation analysis we investigated earlier is attributed to Pearson and assumes that the variables X and Y can be measured on an interval scale. But as we discussed earlier, interval measurement may be impossible or inappropriate in many circumstances. It also assumes that X and Y follow a bivariate normal distribution. Spearman's **rank correlation** coefficient, r_s, is perhaps the most widely used nonparametric measure of correlation and does not have these requirements. It is the ordinary computed correlation using ranks of the data and not the original observations. Interpretation of $r_s = +1$ is that the two samples have the same ranks, while $r_s = -1$ implies that the two samples have exactly inverse ranks, and values of r_s between -1 and $+1$ indicate less than perfect correlation.

To compute r_s, we first rank the data in each sample from smallest to largest. Next we take the difference in each pair of ranks, d_i, square them, sum the squares, and place this value in equation 12.3. Consider the following example of a marketing study of a new white

$$r_3 = 1 - \frac{6\Sigma d_i^2}{n(n^2 - 1)} \qquad \boxed{12.3}$$

grapefruit variety. A randomly selected sample of husbands and wives was asked to rate the product on a scale from 1 to 20, with 20 signifying excellent. Are the ratings correlated at the 5 percent level of significance? The data are given in table 12.4

TABLE 12.4 Husband and Wife Rankings of White Grapefruit for Flavor

Couple	Husband's Rating, X	Wife's Rating, Y	Rank X	Rank Y	d_i
1	12	14	4.5	6	−1.5
2	17	13	10	4.5	5.5
3	9	12	1	3	−2
4	18	20	11	11	0
5	15	17	8	9	−1
6	14	10	7	1	6
7	10	11	2	2	0
8	16	13	9	4.5	4.5
9	13	15	6	7	−1
10	11	18	3	10	−7
11	12	16	4.5	8	−3.5

$$r_s = 1 - \frac{6[-1.5^2 + 5.5^2 + (-2)^2 + 0^2 + (-1)^2 + 6^2 + 0^2 + 4.5^2 + (-1)^2 + (-7)^2 + (-3.5)^2]}{11(11^2 - 1)}$$

$$r = 1 - \frac{936}{1{,}320} = 0.291$$

We compare $r_s = 0.291$ with the critical value of *rho* from Appendix table 14, for $n = 11$ which is 0.623 or greater. Since the computed r_s is less, we cannot reject the null hypothesis of *rho* equal to zero. In the context of this problem, ranks of the husbands' and wives' ratings are not correlated.

Exercises

1. A veterinarian treats two groups of calves with the scours with two different drugs. She gives the first group the standard inoculation dosage and the second group a new drug specially formulated for this condition. The number of hours that elapse until the symptoms disappear are recorded as follows. Is the new drug more effective? Use the Mann-Whitney U test with $\alpha = 0.05$.

Standard dose	15, 9, 12, 22, 14, 9, 10, 13
New drug	7, 8, 11, 6, 7, 7, 4, 11, 13, 5

2. The local grain cooperative wants to select the fastest truck line for shipments to the terminal elevator, served by both Roadhog and Freights-R-Us truck lines. Random samples of shipments to the terminal elevator were monitored with the following results for hours to complete shipments. Is there a difference in the performance of the two truck lines at the 5 percent level of significance? Use the Mann-Whitney U test.

Roadhog	10, 8, 13, 15, 17
Freights-R-Us	14, 9, 12, 14, 11, 8, 13, 20, 16

3. A peach canner puts halves in 16-ounce cans. The manager of the plant sets the machinery to put an average of 16.05 ounces in the cans. The quality control staff samples fifteen cans each hour to determine if the packing process is set properly. The last hour, a series of cans was selected and each can was weighed and marked H if it was heavier than the mean weight and L if it was lighter than the mean weight. Use the runs test on the sample with $\alpha = 0.05$ to see if the runs were generated randomly.

   $$H\,L\,H\,L\,L\,L\,H\,H\,L\,H\,H\,L\,H\,L\,H$$

4. A high school agricultural science teacher decides to test students' learning from two teaching methods on judging livestock. The first method is the standard procedure in which students are taught in the laboratory with live animals. The second method uses a series of videos narrated by an expert judge. The teacher divides his class into two groups, matching each student in the first group with another in the second who has similar skills and aptitudes, and exposes the first group to the standard procedure and the second to the video approach. After the teaching segment is completed, students are given an identical test. Their scores are recorded in the following table. Use the sign test with a 5 percent level of significance to see if the video approach is better.

Standard	Video
66	68
65	67
71	74
72	66
83	80
74	77
92	96
81	90
83	80
76	72
87	88
95	95
90	98
85	95
89	96
75	73
93	96
79	88
88	92
82	78
73	85

5. The residuals from the multiple regression problem in exercise 13 of Chapter 11 are reproduced in the following table. Use the runs test with a 5 percent level of significance to see if the plus and minus values are distributed randomly.

Residual Output

Observation	*Predicted Corn Seed*	*Residuals*
1	236.1417945	−1.14179
2	247.190609	−2.19061
3	258.8335658	1.166434
4	270.4765225	−0.47652
5	252.7150163	2.284984
6	263.7638308	1.236169
7	276.0009298	−1.00093
8	241.6662018	−1.6662
9	231.2115295	1.78847

6. Following are historical data on the price of coffee and the price of tea. Use Spearman's rank correlation procedure to determine whether the two price series are correlated with $\alpha = 0.05$.

Coffee Price	Tea Price
20	32
17	31
14	27
15	30
18	32
19	33
16	30

7. Check the following Production and Price Indexes to determine whether these series are correlated using Spearman's rank correlation procedure with $\alpha = 0.05$.

Production Index	Price Index
99	149
98	128
102	90
96	68
96	72
93	90
91	109
94	114
106	122
103	97

CHAPTER

13

Index Numbers

In a preceding chapter, we used index numbers to depict seasonal variation in time series data. We now want to develop a measure that summarizes the characteristics of large amounts of data. We can use index numbers for this purpose because they relate the values of a dependent variable, such as sales or price, to an independent variable, such as time or income. Although index numbers are used in many areas of the economy, their main application is to describe business and economic activity, such as changes in price, production, wages, and employment over a period of time. There are both simple and composite indexes. A **simple index** expresses the relationship between two numbers, with one of them used as a base. Simple indexes may be percentage relatives and can be used, for example, to show the percent change in the price of labor from one period to another. A **composite index** combines information from several sets of data into one index. For example, the index of prices received by farmers combines several farm product price data sets into a single index that expresses how all farm prices behave over time.

Problems in Index Construction

The most prevalent problems in index number construction are:

1. Selecting the items to be included,
2. Choosing a base period,
3. Determining the mathematical procedure, and
4. Identifying the weights to be used.

Items to Include

Most of the popular price indexes are published by government agencies or large, private concerns and are used in many different ways. Thus, there is no single purpose for price indexes. However, we can attempt to answer meaningful questions with any index, and the major purposes for which we use it determine the items that we include in its construction. Consider the Consumer Price Index, which, for example, most forecasters consider as a general indicator of economic well-being in the economy. But it has other uses: the wage levels of various workers and retirees are tied to it, it is watched by monetary authorities as an indicator of inflation, and so on. This index has several limitations when applied to so many purposes. Nevertheless, its general purpose is to record the average movement of certain prices over time. It is this purpose that indicates what data series will and will not be included in the index, and what is included determines its limitations as well as its strengths. Thus, we need to be certain that we understand what an index measures so we do not make generalizations that the numbers will not support.

How we select items for inclusion in an index depends upon how important they are. Thus, we cannot use random sampling procedures because each good or service cannot be considered as representative as any other. Rather, we include practically all of the important items, and use these to represent the population. After we decide which items to include in the index, then we may use advanced sampling procedures to select the specific items for the index.

The Base Period

The month, year, or series of years from which the index measures changes is called the base period. We always set the index number for the base period at 100 and express values for other time periods as percentages of the base period. Thus, we want to choose a "normal" period for the base. Of course, we may find it difficult to define a normal period for most economic variables, but we do not want to select some period that represents a peak or trough for the series as it fluctuates through time. To construct such an index is not difficult mathematically, but it causes distortions since comparisons made with it are based on abnormal times of peaks or troughs.

Most U.S. government indexes employ a period of years as the base since this provides an averaging effect on year-to-year variations. Also, we want the base period to be fairly close to the present, i.e., we prefer a base of 1990–1994 to 1910–1914. When base periods represent much earlier times, we tend to forget what the prevailing economic conditions were at those times, and our comparisons with such periods tend to lose significance and become difficult to interpret because of technological and social change. Thus, most government agencies shift their base periods every decade or so. Also, if possible, we want to select a common base period that will be used in the construction of other closely affiliated indexes to facilitate comparisons.

Basic Construction Techniques

Most composite indexes are of the aggregative type, while the majority of simple indexes are an average of relatives. In a simple index, the first step involves computing the relative for a commodity by dividing its value in the current (or nonbase) year by the base year value. If several commodities are involved, then we average the relatives to obtain the overall index and may also apply weights as part of the averaging process. In an aggregative type of index, we sum values in a nonbase year and compare them to a summation for the base year, and again we may apply weights during the summation (or aggregation).

Weights

We use weighting schemes because the items included in an index generally do not have the same importance. And in some indexes, such as price indexes, the relative importance of commodities changes over time because of changing market conditions. This causes a problem in index number construction for which we have no satisfactory solution. We generally use quantities as weights in price indexes and are concerned with whether the quantities are from the base period or the current period, or represent some average of the two. The set of quantities we choose for weights influences the value of the index. And in cases in which changing market conditions influence demand for certain commodities, we mute the change if we use base period weights. If we select current period weights, we cause problems with index values for earlier years when demand has not changed, etc.

It is difficult to choose weights that reflect changes in the quality of items especially when there is technological change. Consider a price index that includes personal computers. While prices have moderated to some extent, the original personal computer that IBM marketed in the early 1980s had a price close to its latest version, which has a processor more than 100 times as fast as the original, as well as many recently invented components. Thus, a price index that includes this commodity greatly understates the price level change.

Index Construction

After we select a base period for an index, we choose which kind to construct—simple, relative, weighted, aggregative, etc., and, if weighted, the type of weights. Our final choices depend upon the type of data available to us. Accurate, timely data is a must.

Simple Price Indexes

Because prices play a major role in economic activity, we are generally interested in the construction of price indexes. Most simple indexes describe changes either in a single commodity or in a set of commodities. We use the simple price relative to describe changes in one commodity.

Simple Price Relative

$$I_0 = [p_0/p_0] \times 100 = 100$$
$$I_n = [p_n/p_0] \times 100.$$

The first term in the preceding equation shows that our index number for the base year is 100, but the index number for the nonbase year, I_n, is a percentage increase or decrease in the commodity price from the base period. For example, suppose the national farm price of wheat is $3.24 per bushel in 1998, $3.26 in 1999, and $3.45 in 2000. If we select 1998 as the base year, we construct simple price relatives for the three years as follows:

$$\text{1998 price relative: } [p_0/p_0] \times 100 = [3.24/3.24] \times 100 = 100.0$$
$$\text{1999 price relative: } [p_1/p_0] \times 100 = [3.26/3.24] \times 100 = 100.6$$
$$\text{2000 price relative: } [p_2/p_0] \times 100 = [3.45/3.24] \times 100 = 106.5.$$

Thus, in 1999, farm wheat price is only 0.6 percent higher than it was in 1998, but in 2000, it increases 6.5 percent above the price of wheat in 1998.

If we want to compute an index for the national farm price of grains, we use a simple aggregative price index with the data from the 1998 and 2000 price columns of table 13.1.

Simple Aggregate Price Index

$$I_n = [\Sigma p_n/\Sigma p_0] \times 100$$
$$I_n = [16.65/15.13] \times 100 = 110.$$

The index of 110 states that national farm grain prices increased 10 percent in 2000 over the 1998 base. This index fails to account for the units in which the prices are quoted and does not consider the relative importance of the various types of grains to the industry.

If we change to an index of relatives, we can overcome the first objection.

TABLE 13.1 National Grain Prices and Calculations for Aggregate Price Index

Grain	Units	1998 Price, p_0	2000 Price, p_2	Price Relatives, p_2/p_0
Wheat	Bushel	$ 3.24	$ 3.45	1.06
Rice	Hundredweight	$ 5.89	$ 6.78	1.15
Corn	Bushel	$ 2.07	$ 2.26	1.09
Sorghum	Bushel	$ 1.89	$ 2.13	1.13
Barley	Bushel	$ 2.04	$ 2.03	1.00
Total		$15.13	$16.65	5.43

Simple Average of Price Relatives

$$I_n = \Sigma(p_n/p_0)/N \times 100$$

For the data in table 13.1, we get

$$I_2 = 5.43/5 \times 100 = 108.6$$

Weighted Price Indexes

With weighted indexes, we account for the relative importance of each commodity in the index. Thus, for the national farm grain price example, we may construct an index that considers the importance of each grain in the industry. With a weighted average of price relatives index, we use expenditure weights, where expenditure is price times quantity. Our index has two different forms, depending upon whether we use base year or current year expenditures as weights. We may compute both based on the data from table 13.2.

Weighted Average of Price Relatives with Base Year Weights

$$I_n = \frac{\Sigma \frac{p_n}{p_0}(p_0 q_0)}{\Sigma p_0 q_0} \times 100 \qquad \boxed{13.1}$$

To compute the expression in the numerator of the index in equation 13.1, we need to multiply the price relatives from the last column of table 13.1 by the expenditures from the next to last column of table 13.2 and sum the result (table

TABLE 13.2 National Grain Price and Production Data and Computation of Expenditure Weights

	Prices			Quantities		Expenditures	
Grain	1998, p_0	2000, p_2	Units	1998, q_0	2000, q_2	p_0q_0	p_2q_2
Wheat	$3.24	$3.45	Million bushels	2,467	2,321	$ 7,993.08	$ 8,007.45
Rice	$5.89	$6.78	Million hundredweight	179.7	197.8	$ 1,058.43	$ 1,341.08
Corn	$2.07	$2.26	Million bushels	9,477	10,103	$19,617.39	$22,832.78
Sorghum	$1.89	$2.13	Million bushels	875	649	$ 1,653.75	$ 1,382.37
Barley	$2.04	$2.03	Million bushels	455	375	$ 928.20	$ 761.25
Total						$31,250.85	$34,324.93

13.3). The term in the denominator is the total for the next-to-last column of table 13.2. Thus, our index is:

$$I_n = \frac{\Sigma \frac{p_n}{p_0}(p_0 q_0)}{\Sigma p_0 q_0} \times 100 = \frac{33{,}934.94}{31{,}250.85} \times 100 = 108.6$$

Weighted Average of Price Relatives with Current Year Weights

$$I_n = \frac{\Sigma \frac{p_n}{p_0}(p_n q_n)}{\Sigma p_n q_n} \times 100 \tag{13.2}$$

This index uses the current year expenditures as weights (equation 13.2). For the example data on farm grain price weighted index, we multiply the price relatives in the last column of table 13.1 by the current year expenditures from the last column of table 13.2 to get the weighted price relatives in the last column of table 13.3. The sum of weighted price relatives is the numerator in the index number formula. The denominator is the sum of the expenditures in the current year—the total of the last column in table 13.2. We put the numbers together to get the index:

$$I_n = \frac{\Sigma \frac{p_n}{p_0}(p_n q_n)}{\Sigma p_n q_n} \times 100 = \frac{37{,}314.15}{34{,}324.93} \times 100 = 108.7$$

This weighted relatives index is somewhat closer to the price relative for corn because the expenditure on corn increased in the current year, giving it an even greater influence in the overall index.

Weighted Aggregative Price Index with Base Year Weights: The Laspeyeres Index

$$I_n = \frac{\Sigma p_n q_0}{\Sigma p_0 q_0} \times 100 \tag{13.3}$$

The Laspeyeres index (equation 13.3) weighs each price directly by the quantity consumed in the base year, as in table 13.4. To compute the Laspeyeres index, we total the middle column of table 13.4 and divide it by the total of the next-to-last column of table 13.2. Notice that this index

$$I_n = \frac{\Sigma p_n q_0}{\Sigma p_0 q_0} \times 100 = \frac{33{,}934.94}{31{,}250.85} \times 100 = 108.6$$

TABLE 13.3 Weighted Price Relatives with Base and Current Year Weights

Grain	Base Year Weights $(p_2/p_0)(p_0q_0)$	Current Year Weights $(p_2/p_0)(p_2q_2)$
Wheat	8,511.15	8,526.45
Rice	1,218.37	1,543.73
Corn	21,418.02	24,928.54
Sorghum	1,863.75	1,557.91
Barley	923.65	757.52
Total	33,934.94	37,314.15

provides the same answer as the weighted average of relatives price index using base year weights, and it should since the two are algebraically equivalent.

Weighted Aggregative Price Index with Current Year Weights: The Paasche Index

$$I_n = \frac{\Sigma p_n q_n}{\Sigma p_0 q_n} \times 100 \qquad \boxed{13.4}$$

When we take the current year expenditures and divide by the base year prices weighted with quantities from the current year, we have the Paasche price index (equation 13.4). For the farm grain price data, we divide the total expenditures for the current year from the last column of table 13.2 by the data from the last column of table 13.4 to compute the index. The number we

$$I_n = \frac{\Sigma p_n q_n}{\Sigma p_0 q_n} \times 100 = \frac{34{,}324.93}{31{,}589.90} \times 100 = 108.7$$

TABLE 13.4 Weighted Aggregative Prices with Base and Current Year Weights

Grain	p_2q_0	p_0q_2
Wheat	$ 8,511.15	$ 7,520.04
Rice	$ 1,218.37	$ 1,165.04
Corn	$21,418.02	$20,913.21
Sorghum	$ 1,863.75	$ 1,226.61
Barley	$ 923.65	$ 765.00
Total	$33,934.94	$31,589.90

obtain is the same as for the weighted average of price relatives index using current year weights since the two expressions are equivalent algebraically.

There are also quantity and expenditure indexes in addition to the price indexes we have discussed, but the basic construction techniques are the same. Only the data are different.

Exercises

1. Suppose the market-basket values in the following table were observed in 1999 and 2000.

	1999		2000	
Item	**Price**	**Quantity**	**Price**	**Quantity**
Milk	$1.25/quart	25 quarts	$1.25/quart	20 quarts
Bread	$1.15/loaf	30 loaves	$1.25/loaf	35 loaves
Eggs	$0.90/dozen	10 dozen	$0.80/dozen	15 dozen

   a. Construct price relatives for each commodity.
   b. Construct a simple aggregate price index.
   c. Construct a simple average of price relatives index.
   d. Construct a weighted average of price relatives index using both base year and given year weights.
   e. Construct a Laspeyeres and a Paasche price index.

2. Compute a weighted average of price relatives index using base year weights for the following apple data.

	Price ($/bushel)		Production (million bushels)	
	Fuji	**Macintosh**	**Fuji**	**Macintosh**
1998	$2.45	$2.15	1.28	1.40
1999	$2.52	$2.23	1.32	1.47
2000	$2.60	$2.41	1.31	1.52

3. Your company's gross sales were $12 million in 1998 and $18 million in 2000. It uses the following index as a price deflator. By what percentage did real gross sales increase from 1998 to 2000?

Price	Index
1998	100
1999	123
2000	150

4. The following index of farm labor use has a base year of 1989 = 100. Shift the index so that it has a new base of 1999 = 100.

Year	Index of Farm Labor Use
1985	110
1986	106
1987	102
1988	102
1989	100
1990	95
1991	95
1992	91
1993	85
1994	84
1995	86
1996	85
1997	85
1998	87
1999	82
2000	81

Appendix: Statistical Tables

List of Appendix Tables

APPENDIX TABLE 1 Random Numbers

345769	953810	627280	423578	353511	899906	827008
549075	004410	059309	271243	403382	248735	972383
423480	950812	197145	556566	655917	046169	363201
554518	514280	950974	482196	058868	474936	724289
797165	670995	791954	188521	950156	086813	033365
062730	163375	602168	908350	360861	152201	966097
356756	519371	679389	371912	502903	936741	636775
700770	781547	916968	136999	801855	605975	295802
279584	733750	487151	116069	274869	416181	610911
862434	481154	391464	021094	761599	474456	582253
199585	167701	170778	934765	761328	275799	323046
048736	514507	977406	158840	846761	198016	933522
815218	609732	629295	517386	824505	676788	304971
643021	527212	492869	261844	914505	354436	355772
164332	245407	517804	422658	751712	583087	286872
174303	085157	308590	535846	503131	266915	465641
136325	414066	452293	649359	844625	674828	953396
117780	407444	426115	108970	621527	601599	652376
435697	245510	946158	934221	824917	509832	362638
912252	579474	848845	824321	049853	151126	052643
754438	658573	717914	040054	630638	264060	594641
322053	924909	048177	957012	801464	833319	978384
897199	125506	708669	408374	737887	906201	599469
046637	642050	435779	502427	027842	515775	811203
979346	721653	260190	842505	797017	157497	179041
202312	011976	373248	374293	802292	646914	171322
354014	356787	511271	904434	068589	329862	829316
682909	809290	793392	098004	120575	469925	112743
897690	572456	871574	465543	486529	507767	608677
139029	160636	417690	191242	625269	104858	020808
341447	723998	905614	519309	926345	240082	395043
415603	129727	894956	780924	227496	134056	023014
014881	496311	750082	707823	738906	157591	072396
827235	783798	324650	485324	568156	098331	768720
261607	730824	341940	259028	253973	145183	658110
527920	834376	972906	627959	654790	342497	593779

APPENDIX TABLE 1 Random Numbers (continued)

835431	206253	467521	029822	700399	554652	450184
512651	743206	118787	587401	921517	015407	206860
376187	189133	154812	828785	667020	998697	579598
092530	869028	483691	165063	847894	041617	762973
238036	016856	290105	538530	079931	412195	838814
308168	717698	919814	092230	215657	469994	805803
773429	915639	900911	276895	149505	540379	224349
171626	601259	009905	572567	441960	299704	313987
180570	665625	424048	713009	830314	664642	521021
558715	965963	494210	875287	488595	898691	713010
345067	361180	989224	138905	355519	045847	746266
583819	310956	174728	099164	118461	758000	496302
615026	599459	722322	555090	572720	826686	456517
812358	389535	166779	441968	105639	632418	340890
784592	003651	279275	055646	341897	510689	026160
094619	636747	934082	787345	772825	603866	565688
450908	919891	157771	114333	710179	062848	615156
593546	728768	984323	290410	970562	906724	315005
873778	491131	209695	604075	783895	862911	772026
965705	317845	169619	921361	315606	990029	745251
311163	943589	540958	556212	760508	129963	236556
454554	284761	269019	924179	670780	389869	519229
124330	819763	596075	064570	495169	030185	866211
920765	122124	423205	596357	469969	072245	359269
183002	540547	312909	389818	464023	768381	377241
600135	865974	929756	162716	415598	878513	994633
235787	023117	895285	027055	943962	381112	530492
953379	655834	283102	836259	437761	391976	940853
009658	521970	537626	806052	715247	808585	252503
176570	849057	387097	311529	893745	450267	182626
747456	304530	931013	678688	270736	355032	400713
486876	631985	368395	154273	959983	672523	210456
987193	268135	867829	025419	301168	409545	131960
358155	950977	170562	246987	884126	785621	467942
021394	182615	049084	942153	278313	872709	693590
735047	428941	630704	893281	716045	267529	427605

APPENDIX TABLE 2 Binomial Probabilities

n	x	p										
		0.05	0.1	0.2	0.3	0.4	0.5	0.6	0.7	0.8	0.9	0.95
2	0	0.902	0.810	0.640	0.490	0.360	0.250	0.160	0.090	0.040	0.010	0.002
	1	0.095	0.180	0.320	0.420	0.480	0.500	0.480	0.420	0.320	0.180	0.095
	2	0.002	0.010	0.040	0.090	0.160	0.250	0.360	0.490	0.640	0.810	0.902
3	0	0.857	0.729	0.512	0.343	0.216	0.125	0.064	0.027	0.008	0.001	
	1	0.135	0.243	0.384	0.441	0.432	0.375	0.288	0.189	0.096	0.027	
	2	0.007	0.027	0.096	0.189	0.288	0.375	0.432	0.441	0.384	0.243	0.135
	3		0.001	0.008	0.027	0.064	0.125	0.216	0.343	0.512	0.729	0.857
4	0	0.815	0.656	0.410	0.240	0.130	0.062	0.026	0.008	0.002		
	1	0.171	0.292	0.410	0.412	0.346	0.250	0.154	0.076	0.026	0.004	
	2	0.014	0.049	0.154	0.265	0.346	0.375	0.346	0.265	0.154	0.049	0.014
	3		0.004	0.026	0.076	0.154	0.250	0.346	0.412	0.410	0.292	0.171
	4			0.002	0.008	0.026	0.062	0.130	0.240	0.410	0.656	0.815
5	0	0.774	0.590	0.328	0.168	0.078	0.031	0.010	0.002			
	1	0.204	0.328	0.410	0.360	0.259	0.156	0.077	0.028	0.006		
	2	0.021	0.073	0.205	0.309	0.346	0.312	0.230	0.132	0.051	0.008	0.001
	3	0.001	0.008	0.051	0.132	0.230	0.312	0.346	0.309	0.205	0.073	0.021
	4			0.006	0.028	0.077	0.156	0.259	0.360	0.410	0.328	0.204
	5				0.002	0.010	0.031	0.078	0.168	0.328	0.590	0.774
6	0	0.735	0.531	0.262	0.118	0.047	0.016	0.004	0.001			
	1	0.232	0.354	0.393	0.303	0.187	0.094	0.037	0.010	0.002		
	2	0.031	0.098	0.246	0.324	0.311	0.234	0.138	0.060	0.015	0.001	
	3	0.002	0.015	0.082	0.185	0.276	0.312	0.276	0.185	0.082	0.015	0.002
	4		0.001	0.015	0.060	0.138	0.234	0.311	0.324	0.246	0.098	0.031
	5			0.002	0.010	0.037	0.094	0.187	0.303	0.393	0.354	0.232
	6				0.001	0.004	0.016	0.047	0.118	0.262	0.531	0.735

APPENDIX TABLE 2 Binomial Probabilities (continued)

n	*x*	*p*										
		0.05	0.1	0.2	0.3	0.4	0.5	0.6	0.7	0.8	0.9	0.95
7	0	0.698	0.478	0.210	0.082	0.028	0.008	0.002				
	1	0.257	0.372	0.367	0.247	0.131	0.055	0.017	0.004			
	2	0.041	0.124	0.275	0.318	0.261	0.164	0.077	0.025	0.004		
	3	0.004	0.023	0.115	0.227	0.290	0.273	0.194	0.097	0.029	0.003	
	4		0.003	0.029	0.097	0.194	0.273	0.290	0.227	0.115	0.023	0.004
	5			0.004	0.025	0.077	0.164	0.261	0.318	0.275	0.124	0.041
	6				0.004	0.017	0.055	0.131	0.247	0.367	0.372	0.257
	7					0.002	0.008	0.028	0.082	0.210	0.478	0.698
8	0	0.663	0.430	0.168	0.058	0.017	0.004	0.001				
	1	0.279	0.383	0.336	0.198	0.090	0.031	0.008	0.001			
	2	0.051	0.149	0.294	0.296	0.209	0.109	0.041	0.010	0.001		
	3	0.005	0.033	0.147	0.254	0.279	0.219	0.124	0.047	0.009		
	4		0.005	0.046	0.136	0.232	0.273	0.232	0.116	0.046	0.005	
	5			0.009	0.047	0.124	0.219	0.279	0.254	0.147	0.033	0.005
	6			0.001	0.010	0.041	0.109	0.209	0.296	0.294	0.149	0.051
	7				0.001	0.008	0.031	0.090	0.198	0.336	0.383	0.279
	8					0.001	0.004	0.017	0.058	0.168	0.430	0.663
9	0	0.630	0.387	0.134	0.040	0.010	0.002					
	1	0.299	0.387	0.302	0.156	0.060	0.018	0.004				
	2	0.063	0.172	0.302	0.267	0.161	0.070	0.021	0.004			
	3	0.008	0.045	0.176	0.267	0.251	0.164	0.074	0.021	0.003		
	4	0.001	0.007	0.066	0.172	0.251	0.246	0.167	0.074	0.017	0.001	
	5		0.001	0.017	0.074	0.167	0.246	0.251	0.172	0.066	0.007	0.001
	6			0.003	0.021	0.074	0.164	0.251	0.267	0.176	0.045	0.008
	7				0.004	0.021	0.070	0.161	0.267	0.302	0.172	0.063
	8					0.004	0.018	0.060	0.156	0.302	0.387	0.299
	9						0.002	0.010	0.040	0.134	0.387	0.630

APPENDIX TABLE 2 Binomial Probabilities (continued)

n	x	p										
		0.05	0.1	0.2	0.3	0.4	0.5	0.6	0.7	0.8	0.9	0.95
10	0	0.599	0.349	0.107	0.028	0.006	0.001					
	1	0.315	0.387	0.268	0.121	0.040	0.010	0.002				
	2	0.075	0.194	0.302	0.'33	0.121	0.044	0.011	0.001			
	3	0.010	0.057	0.201	0.267	0.215	0.117	0.042	0.009	0.001		
	4	0.001	0.011	0.088	0.200	0.251	0.205	0.111	0.037	0.006		
	5		0.001	0.026	0.103	0.201	0.246	0.201	0.103	0.026	0.001	
	6			0.006	0.037	0.111	0.205	0.251	0.200	0.088	0.011	0.001
	7			0.001	0.009	0.042	0.117	0.215	0.267	0.201	0.057	0.010
	8				0.001	0.011	0.044	0.121	0.233	0.302	0.194	0.075
	9					0.002	0.010	0.040	0.121	0.268	0.387	0.315
	10						0.001	0.006	0.028	0.107	0.349	0.599
11	0	0.569	0.314	0.086	0.020	0.004						
	1	0.329	0.384	0.236	0.093	0.027	0.005	0.001				
	2	0.087	0.213	0.295	0.200	0.089	0.027	0.005	0.001			
	3	0.014	0.071	0.221	0.257	0.177	0.081	0.023	0.004			
	4	0.001	0.016	0.111	0.220	0.236	0.161	0.070	0.017	0.002		
	5		0.002	0.039	0.132	0.221	0.226	0.147	0.057	0.010		
	6			0.010	0.057	0.147	0.226	0.221	0.132	0.039	0.002	
	7			0.002	0.017	0.070	0.161	0.236	0.220	0.111	0.016	0.001
	8				0.004	0.023	0.081	0.177	0.257	0.221	0.071	0.014
	9				0.001	0.005	0.027	0.089	0.200	0.295	0.213	0.087
	10					0.001	0.005	0.027	0.093	0.236	0.384	0.329
	11							0.004	0.020	0.086	0.314	0.569

APPENDIX TABLE 2 Binomial Probabilities (continued)

n	x	p										
		0.05	0.1	0.2	0.3	0.4	0.5	0.6	0.7	0.8	0.9	0.95
12	0	0.540	0.282	0.069	0.014	0.002						
	1	0.341	0.377	0.206	0.071	0.017	0.003					
	2	0.099	0.230	0.283	0.168	0.064	0.016	0.002				
	3	0.017	0.085	0.236	0.240	0.142	0.054	0.012	0.001			
	4	0.002	0.021	0.133	0.231	0.213	0.121	0.042	0.008	0.001		
	5		0.004	0.053	0.158	0.227	0.193	0.101	0.029	0.003		
	6			0.016	0.079	0.177	0.226	0.177	0.079	0.016		
	7			0.003	0.029	0.101	0.193	0.227	0.158	0.053	0.004	
	8			0.001	0.008	0.042	0.121	0.213	0.231	0.133	0.021	0.002
	9				0.001	0.012	0.054	0.142	0.240	0.236	0.085	0.017
	10					0.002	0.016	0.064	0.168	0.283	0.230	0.099
	11						0.003	0.017	0.071	0.206	0.377	0.341
	12							0.002	0.014	0.069	0.282	0.540
13	0	0.513	0.254	0.055	0.010	0.001						
	1	0.351	0.367	0.179	0.054	0.011	0.002					
	2	0.111	0.245	0.268	0.139	0.045	0.010	0.001				
	3	0.021	0.100	0.246	0.218	0.111	0.035	0.006	0.001			
	4	0.003	0.028	0.154	0.234	0.184	0.087	0.024	0.003			
	5		0.006	0.069	0.180	0.221	0.157	0.066	0.014	0.001		
	6		0.001	0.023	0.103	0.197	0.209	0.131	0.044	0.006		
	7			0.006	0.044	0.131	0.209	0.197	0.103	0.023	0.001	
	8			0.001	0.014	0.066	0.157	0.221	0.180	0.069	0.006	
	9				0.003	0.024	0.087	0.184	0.234	0.154	0.028	0.003
	10				0.001	0.006	0.035	0.111	0.218	0.246	0.100	0.021
	11					0.001	0.010	0.045	0.139	0.268	0.245	0.111
	12						0.002	0.011	0.054	0.179	0.367	0.351
	13							0.001	0.010	0.055	0.254	0.513

APPENDIX TABLE 2 Binomial Probabilities (continued)

n	*x*	*p*										
		0.05	0.1	0.2	0.3	0.4	0.5	0.6	0.7	0.8	0.9	0.95
14	0	0.488	0.229	0.044	0.007	0.001						
	1	0.359	0.356	0.154	0.041	0.007	0.001					
	2	0.123	0.257	0.250	0.113	0.032	0.006	0.001				
	3	0.026	0.114	0.250	0.194	0.085	0.022	0.003				
	4	0.004	0.035	0.172	0.229	0.155	0.061	0.014	0.001			
	5		0.008	0.086	0.196	0.207	0.122	0.041	0.007			
	6		0.001	0.032	0.126	0.207	0.183	0.092	0.023	0.002		
	7			0.009	0.062	0.157	0.209	0.157	0.062	0.009		
	8			0.002	0.023	0.092	0.183	0.207	0.126	0.032	0.001	
	9				0.007	0.041	0.122	0.207	0.196	0.086	0.008	
	10				0.001	0.014	0.061	0.155	0.229	0.172	0.035	0.004
	11					0.003	0.022	0.085	0.194	0.250	0.114	0.026
	12					0.001	0.006	0.032	0.113	0.250	0.257	0.123
	13						0.001	0.007	0.041	0.154	0.356	0.359
	14							0.001	0.007	0.044	0.229	0.488
15	0	0.463	0.206	0.035	0.005							
	1	0.366	0.343	0.132	0.031	0.005						
	2	0.135	0.267	0.231	0.092	0.022	0.003					
	3	0.031	0.129	0.250	0.170	0.063	0.014	0.002				
	4	0.005	0.043	0.188	0.219	0.127	0.042	0.007	0.001			
	5	0.001	0.010	0.103	0.206	0.186	0.092	0.024	0.003			
	6		0.002	0.043	0.147	0.207	0.153	0.061	0.012	0.001		
	7			0.014	0.081	0.177	0.196	0.118	0.035	0.003		
	8			0.003	0.035	0.118	0.196	0.177	0.081	0.014		
	9			0.001	0.012	0.061	0.153	0.207	0.147	0.043	0.002	
	10				0.003	0.024	0.092	0.186	0.206	0.103	0.010	0.001
	11				0.001	0.007	0.042	0.127	0.219	0.188	0.043	0.005
	12					0.002	0.014	0.063	0.170	0.250	0.129	0.031
	13						0.003	0.022	0.092	0.231	0.267	0.135
	14							0.005	0.031	0.132	0.343	0.366
	15								0.005	0.035	0.206	0.463

APPENDIX TABLE 3 Binomial Coefficients, $_nC_r$

n	r 0	1	2	3	4	5	6	7	8
0	1								
1	1	1							
2	1	2	1						
3	1	3	3	1					
4	1	4	6	4	1				
5	1	5	10	10	5	1			
6	1	6	15	20	15	6	1		
7	1	7	21	35	35	21	7	1	
8	1	8	28	56	70	56	28	8	1
9	1	9	36	84	126	126	84	36	9
10	1	10	45	120	210	252	210	120	45
11	1	11	55	165	330	462	462	330	165
12	1	12	66	220	495	792	924	792	495
13	1	13	78	286	715	1,287	1,716	1,716	1,287
14	1	14	91	364	1,001	2,002	3,003	3,432	3,003
15	1	15	105	455	1,365	3,003	5,005	6,435	6,435

For coefficients missing from the table, use the relation $_nC_r = {_nC_{n-r}}$
For example $_{12}C_{11} = {_{12}C_1} = 12$

APPENDIX TABLE 4 Factorials

n	n!
0	1
1	1
2	2
3	6
4	24
5	120
6	720
7	5,040
8	40,320
9	362,880
10	3,628,800
11	39,916,800
12	479,001,600
13	6,227,020,800
14	87,178,291,200
15	1,307,674,368,000

APPENDIX TABLE 5 Values of e^{-x}

x	e^{-x}	x	e^{-x}	x	e^{-x}	x	e^{-x}
0.0	1.000	2.5	0.0821	5.0	0.00674	7.5	0.000553
0.1	0.9048	2.6	0.0743	5.1	0.00610	7.6	0.000500
0.2	0.8187	2.7	0.0672	5.2	0.00552	7.7	0.000453
0.3	0.7408	2.8	0.0608	5.3	0.00499	7.8	0.000410
0.4	0.6703	2.9	0.0550	5.4	0.00452	7.9	0.000371
0.5	0.6065	3.0	0.0498	5.5	0.00409	8.0	0.000335
0.6	0.5488	3.1	0.0450	5.6	0.00370	8.1	0.000304
0.7	0.4966	3.2	0.0408	5.7	0.00335	8.2	0.000275
0.8	0.4493	3.3	0.0369	5.8	0.00303	8.3	0.000249
0.9	0.4066	3.4	0.0334	5.9	0.00274	8.4	0.000225
1.0	0.3679	3.5	0.0302	6.0	0.00248	8.5	0.000203
1.1	0.3329	3.6	0.0273	6.1	0.00224	8.6	0.000184
1.2	0.3012	3.7	0.0247	6.2	0.00203	8.7	0.000167
1.3	0.2725	3.8	0.0224	6.3	0.00184	8.8	0.000151
1.4	0.2466	3.9	0.0202	6.4	0.00166	8.9	0.000136
1.5	0.2231	4.0	0.0183	6.5	0.00150	9.0	0.000123
1.6	0.2019	4.1	0.0166	6.6	0.00136	9.1	0.000112
1.7	0.1827	4.2	0.0150	6.7	0.00123	9.2	0.000101
1.8	0.1653	4.3	0.0136	6.8	0.00111	9.3	0.000091
1.9	0.1496	4.4	0.0123	6.9	0.00101	9.4	0.000083
2.0	0.1353	4.5	0.0111	7.0	0.00091	9.5	0.000075
2.1	0.1224	4.6	0.0101	7.1	0.00082	9.6	0.000068
2.2	0.1108	4.7	0.0091	7.2	0.00075	9.7	0.000061
2.3	0.1003	4.8	0.0082	7.3	0.00068	9.8	0.000056
2.4	0.0907	4.9	0.0074	7.4	0.00061	9.9	0.000050

APPENDIX TABLE 6 Poisson Distribution for Selected Values of mu

Values in the table are for the function:

$$P(X) = \frac{\mu^x e^{-\mu}}{x!}$$

$P(X)$ for specified values of μ

	μ									
x	**0.10**	**0.20**	**0.30**	**0.40**	**0.50**	**0.60**	**0.70**	**0.80**	**0.90**	**1.00**
0	.9048	.8187	.7408	.6703	.6065	.5488	.4966	.4493	.4066	.3679
1	.0904	.1637	.2222	.2681	.3033	.3293	.3476	.3595	.3659	.3679
2	.0045	.0164	.0333	.0536	.0758	.0988	.1217	.1438	.1647	.1839
3	.0001	.0011	.0033	.0072	.0126	.0198	.0284	.0383	.0494	.0613
4	.0000	.0001	.0003	.0007	.0016	.0030	.0050	.0077	.0111	.0153
5	.0000	.0000	.0000	.0001	.0002	.0004	.0007	.0012	.0020	.0031
6	.0000	.0000	.0000	.0000	.0000	.0000	.0001	.0002	.0003	.0005
7	.0000	.0000	.0000	.0000	.0000	.0000	.0000	.0000	.0000	.0001

	μ									
x	**1.10**	**1.20**	**1.30**	**1.40**	**1.50**	**1.60**	**1.70**	**1.80**	**1.90**	**2.00**
0	.3329	.3012	.2725	.2466	.2231	.2019	.1827	.1653	.1496	.1353
1	.3662	.3614	.3543	.3452	.3347	.3230	.3106	.2975	.2842	.2707
2	.2014	.2169	.2303	.2417	.2510	.2584	.2640	.2678	.2700	.2707
3	.0738	.0867	.0998	.1128	.1255	.1378	.1496	.1607	.1710	.1804
4	.0203	.0260	.0324	.0395	.0471	.0551	.0636	.0723	.0812	.0902
5	.0045	.0062	.0084	.0111	.0141	.0176	.0216	.0260	.0309	.0361
6	.0008	.0012	.0018	.0026	.0035	.0047	.0061	.0078	.0098	.0120
7	.0001	.0002	.0003	.0005	.0008	.0011	.0015	.0020	.0027	.0034
8	.0000	.0000	.0001	.0001	.0001	.0002	.0003	.0005	.0006	.0009
9	.0000	.0000	.0000	.0000	.0000	.0000	.0001	.0001	.0001	.0002

APPENDIX TABLE 6 Poisson Distribution for Selected Values of mu (continued)

	μ									
x	**2.10**	**2.20**	**2.30**	**2.40**	**2.50**	**2.60**	**2.70**	**2.80**	**2.90**	**3.00**
0	.1125	.1108	.1103	.0907	.0821	.0743	.0672	.0608	.0550	.0498
1	.2572	.2438	.2306	.2177	.2052	.1931	.1815	.1703	.1596	.1494
2	.2700	.2681	.2652	.2613	.2565	.2510	.2450	.2384	.2314	.2240
3	.1890	.1966	.2033	.2090	.2138	.2176	.2205	.2225	.2237	.2240
4	.0992	.1082	.1169	.1254	.1336	.1414	.1488	.1557	.1622	.1680
5	.0417	.0476	.0538	.0602	.0668	.0735	.0804	.0872	.0940	.1008
6	.0146	.0174	.0206	.0241	.0278	.0319	.0362	.0407	.0455	.0504
7	.0044	.0055	.0068	.0083	.0099	.0118	.0139	.0163	.0188	.0216
8	.0011	.0015	.0019	.0025	.0031	.0038	.0047	.0057	.0068	.0081
9	.0003	.0004	.0005	.0007	.0009	.0011	.0014	.0018	.0022	.0027
10	.0001	.0001	.0001	.0002	.0002	.0003	.0004	.0005	.0006	.0008
11	.0000	.0000	.0000	.0000	.0000	.0001	.0001	.0001	.0002	.0002
12	.0000	.0000	.0000	.0000	.0000	.0000	.0000	.0000	.0000	.0001

	μ									
x	**3.10**	**3.20**	**3.30**	**3.40**	**3.50**	**3.60**	**3.70**	**3.80**	**3.90**	**4.00**
0	.0450	.0408	.0369	.0334	.0302	.0273	.0247	.0224	.0202	.0183
1	.1397	.1304	.1217	.1135	.1057	.0984	.0915	.0850	.0789	.0733
2	.2165	.2087	.2008	.1929	.1850	.1771	.1692	.1615	.1539	.1465
3	.2237	.2226	.2209	.2186	.2158	.2125	.2087	.2046	.2001	.1954
4	.1733	.1781	.1823	.1858	.1888	.1912	.1931	.1944	.1951	.1954
5	.1075	.1140	.1203	.1264	.1322	.1377	.1429	.1477	.1522	.1563
6	.0555	.0608	.0662	.0716	.0771	.0826	.0881	.0936	.0989	.1042
7	.0246	.0278	.0312	.0348	.0385	.0425	.0466	.0508	.0551	.0595
8	.0095	.0111	.0129	.0148	.0169	.0191	.0215	.0241	.0269	.0298
9	.0033	.0040	.0047	.0056	.0066	.0076	.0089	.0102	.0116	.0132
10	.0010	.0013	.0016	.0019	.0023	.0028	.0033	.0039	.0045	.0053
11	.0003	.0004	.0005	.0006	.0007	.0009	.0011	.0013	.0016	.0019
12	.0001	.0001	.0001	.0002	.0002	.0003	.0003	.0004	.0005	.0006
13	.0000	.0000	.0000	.0000	.0001	.0001	.0001	.0001	.0002	.0002
14	.0000	.0000	.0000	.0000	.0000	.0000	.0000	.0000	.0000	.0001

	μ									
x	**4.10**	**4.20**	**4.30**	**4.40**	**4.50**	**4.60**	**4.70**	**4.80**	**4.90**	**5.00**
0	.0166	.0150	.0136	.0123	.0111	.0101	.0091	.0082	.0074	.0067
1	.0679	.0630	.0583	.0540	.0500	.0462	.0427	.0395	.0365	.0337
2	.1393	.1323	.1254	.1188	.1125	.1063	.1005	.0948	.0894	.0842

APPENDIX TABLE 6 Poisson Distribution for Selected Values of mu (continued)

	μ									
x	**4.10**	**4.20**	**4.30**	**4.40**	**4.50**	**4.60**	**4.70**	**4.80**	**4.90**	**5.00**
3	.1904	.1852	.1798	.1743	.1687	.1631	.1574	.1517	.1460	.1404
4	.1951	.1944	.1933	.1917	.1898	.1875	.1849	.1820	.1789	.1755
5	.1600	.1633	.1662	.1687	.1708	.1725	.1738	.1747	.1753	.1755
6	.1093	.1143	.1191	.1237	.1281	.1323	.1362	.1398	.1432	.1462
7	.0640	.0686	.0732	.0778	.0824	.0869	.0914	.0959	.1002	.1044
8	.0328	.0360	.0393	.0428	.0463	.0500	.0537	.0575	.0614	.0653
9	.0150	.0168	.0188	.0209	.0232	.0255	.0281	.0307	.0334	.0363
10	.0061	.0071	.0081	.0092	.0104	.0118	.0132	.0147	.0164	.0181
11	.0023	.0027	.0032	.0037	.0043	.0049	.0056	.0064	.0073	.0082
12	.0008	.0009	.0011	.0013	.0016	.0019	.0022	.0026	.0030	.0034
13	.0002	.0003	.0004	.0005	.0006	.0007	.0008	.0009	.0011	.003
14	.0001	.0001	.0001	.0001	.0002	.0002	.0003	.0003	.0004	.0005
15	.0000	.0000	.0000	.0000	.0001	.0001	.0001	.0001	.0001	.0002

	μ									
x	**5.10**	**5.20**	**5.30**	**5.40**	**5.50**	**5.60**	**5.70**	**5.80**	**5.90**	**6.00**
0	.0061	.0055	.0050	.0045	.0041	.0037	.0033	.0030	.0027	.0025
1	.0311	.0287	.0265	.0244	.0225	.0207	.0191	.0176	.0162	.0149
2	.0793	.0746	.0701	.0659	.0618	.0580	.0544	.0509	.0477	.0446
3	.1348	.1293	.1239	.1185	.1133	.1082	.1033	.0985	.0938	.0892
4	.1719	.1681	.1641	.1600	.1558	.1515	.1472	.1428	.1383	.1339
5	.1753	.1748	.1740	.1728	.1714	.1697	.1678	.1656	.1632	.1606
6	.1490	.1515	.1537	.1555	.1571	.1584	.1594	.1601	.1605	.1606
7	.1086	.1125	.1163	.1200	.1234	.1267	.1298	.1326	.1353	.1377
8	.0692	.0731	.0771	.0810	.0849	.0887	.0925	.0962	.0998	.1033
9	.0392	.0423	.0454	.0486	.0519	.0552	.0586	.0620	.0654	.0688
10	.0200	.0220	.0241	.0262	.0285	.0309	.0334	.0359	.0386	.0413
11	.0093	.0104	.0116	.0129	.0143	.0157	.0173	.0190	.0207	.0225
12	.0039	.0045	.0051	.0058	.0065	.0073	.0082	.0092	.0102	.0113
13	.0015	.0018	.0021	.0024	.0028	.0032	.0036	.0041	.0046	.0052
14	.0006	.0007	.0008	.0009	.0011	.0013	.0015	.0017	.0019	.0022
15	.0002	.0002	.0003	.0003	.0004	.0005	.0006	.0007	.0008	.0009
16	.0001	.0001	.0001	.0001	.0001	.0002	.0002	.0002	.0003	.0003
17	.0000	.0000	.0000	.0000	.0000	.0001	.0001	.0001	.0001	.0001

APPENDIX TABLE 6 Poisson Distribution for Selected Values of mu (continued)

	μ									
x	**6.10**	**6.20**	**6.30**	**6.40**	**6.50**	**6.60**	**6.70**	**6.80**	**6.90**	**7.00**
0	.0022	.0020	.0018	.0017	.0015	.0014	.0012	.0011	.0010	.0009
1	.0137	.0126	.0116	.0106	.0098	.0090	.0082	.0076	.0070	.0064
2	.0417	.0390	.0364	.0340	.0318	.0296	.0276	.0258	.0240	.0223
3	.0848	.0806	.0765	.0726	.0688	.0652	.0617	.0584	.0552	.0521
4	.1294	.1249	.1205	.1161	.1118	.1076	.1034	.0992	.0952	.0912
5	.1579	.1549	.1519	.1487	.1454	.1420	.1385	.1349	.1314	.1277
6	.1605	.1601	.1595	.1586	.1575	.1562	.1546	.1529	.1511	.1490
7	.1399	.1418	.1435	.1450	.1462	.1472	.1480	.1486	.1489	.1490
8	.1066	.1099	.1130	.1160	.1188	.1215	.1240	.1263	.1284	.1304
9	.0723	.0757	.0791	.0825	.0858	.0891	.0923	.0954	.0985	.1014
10	.0441	.0469	.0498	.0528	.0558	.0588	.0618	.0649	.0679	.0710
11	.0244	.0265	.0285	.0307	.0330	.0353	.0377	.0401	.0426	.0452
12	.0124	.0137	.0150	.0164	.0179	.0194	.0210	.0227	.0245	.0263
13	.0058	.0065	.0073	.0081	.0089	.0099	.0108	.0119	.0130	.0142
14	.0025	.0029	.0033	.0037	.0041	.0046	.0052	.0058	.0064	.0071
15	.0010	.0012	.0014	.0016	.0018	.0020	.0023	.0026	.0029	.0033
16	.0004	.0005	.0005	.0006	.0007	.0008	.0010	.0011	.0013	.0014
17	.0001	.0002	.0002	.0002	.0003	.0003	.0004	.0004	.0005	.0006
18	.0000	.0001	.0001	.0001	.0001	.0001	.0001	.0002	.0002	.0002
19	.0000	.0000	.0000	.0000	.0000	.0000	.0001	.0001	.0001	.0001

	μ									
x	**7.10**	**7.20**	**7.30**	**7.40**	**7.50**	**7.60**	**7.70**	**7.80**	**7.90**	**8.00**
0	.0008	.0007	.0007	.0006	.0006	.0005	.0005	.0004	.0004	.0003
1	.0059	.0054	.0049	.0045	.0041	.0038	.0035	.0032	.0029	.0027
2	.0208	.0194	.0180	.0167	.0156	.0145	.0134	.0125	.0116	.0107
3	.0492	.0464	.0438	.0413	.0389	.0366	.0345	.0324	.0305	.0286
4	.0874	.0836	.0799	.0764	.0729	.0696	.0663	.0632	.0602	.0573
5	.1241	.1204	.1167	.1130	.1094	.1057	.1021	.0986	.0951	.0916
6	.1468	.1445	.1420	.1394	.1367	.1339	.1311	.1282	.1252	.1221
7	.1489	.1486	.1481	.1474	.1465	.1454	.1442	.1428	.1413	.1396
8	.1321	.1337	.1351	.1363	.1373	.1381	.1388	.1392	.1395	.1396
9	.1042	.1070	.1096	.1121	.1144	.1167	.1187	.1207	.1224	.1241
10	.0740	.0770	.0800	.0829	.0858	.0887	.0914	.0941	.0967	.0993
11	.0478	.0504	.0531	.0558	.0585	.0613	.0640	.0667	.0695	.0722

APPENDIX TABLE 6 Poisson Distribution for Selected Values of mu (continued)

	μ									
x	7.10	7.20	7.30	7.40	7.50	7.60	7.70	7.80	7.90	8.00
12	.0283	.0303	.0323	.0344	.0366	.0388	.0411	.0434	.0457	.0481
13	.0154	.0168	.0181	.0196	.0211	.0227	.0243	.0260	.0278	.0296
14	.0078	.0086	.0095	.0104	.0113	.0123	.0134	.0145	.0157	.0169
15	.0037	.0041	.0046	.0051	.0057	.0062	.0069	.0075	.0083	.0090
16	.0016	.0019	.0021	.0024	.0026	.0030	.0033	.0037	.0041	.0045
17	.0007	.0008	.0009	.0010	.0012	.0013	.0015	.0017	.0019	.0021
18	.0003	.0003	.0004	.0004	.0005	.0006	.0006	.0007	.0008	.0009
19	.0001	.0001	.0001	.0002	.0002	.0002	.0003	.0003	.0003	.0004
20	.0000	.0000	.0001	.0001	.0001	.0001	.0001	.0001	.0001	.0002
21	.0000	.0000	.0000	.0000	.0000	.0000	.0000	.0000	.0001	.0001

	μ									
x	8.10	8.20	8.30	8.40	8.50	8.60	8.70	8.80	8.90	9.00
0	.0003	.0003	.0002	.0002	.0002	.0002	.0002	.0002	.0001	.0001
1	.0025	.0023	.0021	.0019	.0017	.0016	.0014	.0013	.0012	.0011
2	.0100	.0092	.0086	.0079	.0074	.0068	.0063	.0058	.0054	.0050
3	.0269	.0252	.0237	.0222	.0208	.0195	.0183	.0171	.0160	.0150
4	.0544	.0517	.0491	.0466	.0443	.0420	.0398	.0377	.0357	.0337
5	.0882	.0849	.0816	.0784	.0752	.0722	.0692	.0663	.0635	.0607
6	.1192	.1160	.1128	.1097	.1066	.1034	.1003	.0972	.0941	.0911
7	.1378	.1358	.1338	.1317	.1294	.1271	.1247	.1222	.1197	.1171
8	.1395	.1392	.1388	.1382	.1375	.1366	.1356	.1344	.1332	.1318
9	.1256	.1269	.1280	.1290	.1299	.1306	.1311	.1315	.1317	.1318
10	.1017	.1040	.1063	.1084	.1104	.1123	.1140	.1157	.1172	.1186
11	.0749	.0776	.0802	.0928	.0853	.0878	.0902	.0925	.0948	.0970
12	.0505	.0530	.0555	.0579	.0604	.0629	.0654	.0679	.0703	.0728
13	.0315	.0334	.0354	.0374	.0395	.0416	.0438	.0459	.0481	.0504
14	.0182	.0196	.0210	.0225	.0240	.0256	.0272	.0289	.0306	.0324
15	.0098	.0107	.0116	.0126	.0136	.0147	.0158	.0169	.0182	.0194
16	.0050	.0055	.0060	.0066	.0072	.0079	.0086	.0093	.0101	.0109
17	.0024	.0026	.0029	.0033	.0036	.0040	.0044	.0048	.0053	.0058
18	.0011	.0012	.0014	.0015	.0017	.0019	.0021	.0024	.0026	.0029
19	.0005	.0005	.0006	.0007	.0008	.0009	.0010	.0011	.0012	.0014
20	.0002	.0002	.0002	.0003	.0003	.0004	.0004	.0005	.0005	.0006
21	.0001	.0001	.0001	.0001	.0001	.0002	.0002	.0002	.0002	.0003
22	.0000	.0000	.0000	.0000	.0001	.0001	.0001	.0001	.0001	.0001

APPENDIX TABLE 6 Poisson Distribution for Selected Values of mu (continued)

	μ									
x	**9.10**	**9.20**	**9.30**	**9.40**	**9.50**	**9.60**	**9.70**	**9.80**	**9.90**	**10.00**
0	.0001	.0001	.0001	.0001	.0001	.0001	.0001	.0001	.0001	
1	.0010	.0009	.0009	.0008	.0007	.0007	.0006	.0005	.0005	.0005
2	.0046	.0043	.0040	.0037	.0034	.0031	.0029	.0027	.0025	.0023
3	.0140	.0131	.0123	.0115	.0107	.0100	.0093	.0087	.0081	.0076
4	.0319	.0302	.0285	.0269	.0254	.0240	.0226	.0213	.0201	.0189
5	.0581	.0555	.0530	.0506	.0483	.0460	.0439	.0418	.0398	.0378
6	.0881	.0851	.0822	.0793	.0764	.0736	.0709	.0682	.0656	.0631
7	.1145	.1118	.1091	.1064	.1037	.1010	.0982	.0955	.0928	.0901
8	.1302	.1286	.1269	.1251	.1232	.1212	.1191	.1170	.1148	.1126
9	.1317	.1315	.1311	.1306	.1300	.1293	.1284	.1274	.1263	.1251
10	.1198	.1210	.1219	.1228	.1235	.1241	.1245	.1249	.1250	.1251
11	.0991	.1012	.1031	.1049	.1067	.1083	.1098	.1112	.1125	.1137
12	.0752	.0776	.0799	.0822	.0844	.0866	.0888	.0908	.0928	.0948
13	.0526	.0549	.0572	.0594	.0617	.0640	.0662	.0685	.0707	.0729
14	.0342	.0361	.0380	.0399	.0419	.0439	.0459	.0479	.0500	.0521
15	.0208	.0221	.0235	.0250	.0265	.0281	.0297	.0313	.0330	.0347
16	.0118	.0127	.0137	.0147	.0157	.0168	.0180	.0192	.0204	.0217
17	.0063	.0069	.0075	.0081	.0088	.0095	.0103	.0111	.0119	.0128
18	.0032	.0035	.0039	.0042	.0046	.0051	.0055	.0060	.0065	.0071
19	.0015	.0017	.0019	.0021	.0023	.0026	.0028	.0031	.0034	.0037
20	.0007	.0008	.0009	.0010	.0011	.0012	.0014	.0015	.0017	.0019
21	.0003	.0003	.0004	.0004	.0005	.0006	.0006	.0007	.0008	.0009
22	.0001	.0001	.0002	.0002	.0002	.0002	.0003	.0003	.0004	.0004
23	.0000	.0001	.0001	.0001	.0001	.0001	.0001	.0001	.0002	.0002
24	.0000	.0000	.0000	.0000	.0000	.0000	.0000	.0001	.0001	.0001

APPENDIX TABLE 7 Normal Curve Areas

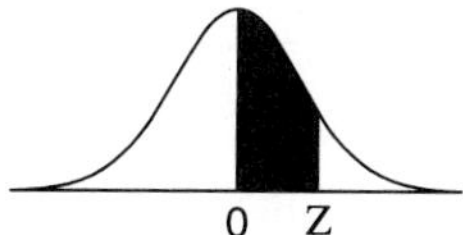

How to use this table: Example 1—Suppose you want a probability for a Z of 1.57. Look under the first column for 1.5 and read across to .07 and find the value .4418, which is bolded so you can see it easily.

Example 2—Suppose you want a Z that cuts off an area in the right tail of the distribution equal to 0.025. The table gives values for areas between the mean and Z. So first subtract 0.025 from 0.5, which represents the area to the right of the mean. Look up the answer in the table body and read off the Z value in the row and column headings. To find the Z for 0.025: .5000–.0250 = .4750. Find .4750 in the body of the table and read off the Z of 1.96. These values are also bolded so you can follow easily.

Z	.00	.01	.02	.03	.04	.05	.06	.07	.08	.09
0.0	.0000	.0040	.0080	.0120	.0160	.0199	.0239	.0279	.0319	.0359
0.1	.0398	.0438	.0478	.0517	.0557	.0596	.0636	.0675	.0714	.0753
0.2	.0793	.0832	.0871	.0910	.0948	.0987	.1026	.1064	.1103	.1141
0.3	.1179	.1217	.1255	.1293	.1331	.1368	.1406	.1443	.1480	.1517
0.4	.1554	.1591	.1628	.1664	.1700	.1736	.1772	.1808	.1844	.1879
0.5	.1915	.1950	.1985	.2019	.2054	.2088	.2123	.2157	.2190	.2224
0.6	.2257	.2291	.2324	.2357	.2389	.2422	.2454	.2486	.2517	.2549
0.7	.2580	.2611	.2642	.2673	.2704	.2734	.2764	.2794	.2823	.2852
0.8	.2881	.2910	.2939	.2967	.2995	.3023	.3051	.3078	.3106	.3133
0.9	.3159	.3186	.3212	.3238	.3264	.3289	.3315	.3340	.3365	.3389
1.0	.3413	.3438	.3461	.3485	.3508	.3531	.3554	.3577	.3599	.3621
1.1	.3643	.3665	.3686	.3708	.3729	.3749	.3770	.3790	.3810	.3830
1.2	.3849	.3869	.3888	.3907	.3925	.3944	.3962	.3980	.3997	.4015
1.3	.4032	.4049	.4066	.4082	.4099	.4115	.4131	.4147	.4162	.4177
1.4	.4192	.4207	.4222	.4236	.4251	.4265	.4279	.2492	.4306	.4319
1.5	.4332	.4345	.4357	.4370	.4382	.4394	.4406	**.4418**	.4430	.4441
1.6	.4452	.4463	.4474	.4484	.4495	.4505	.4515	.4525	.4535	.4545
1.7	.4554	.4564	.4573	.4582	.4591	.4599	.4608	.4616	.4625	.4633
1.8	.4641	.4649	.4656	.4664	.4671	.4678	.4686	.4693	.4699	.4706
1.9	.4713	.4719	.4726	.4732	.4738	.4744	**.4750**	.4756	.4761	.4767
2.0	.4772	.4778	.4783	.4788	.4793	.4798	.4803	.4808	.4812	.4817

APPENDIX TABLE 7 Normal Curve Areas (continued)

Z	.00	.01	.02	.03	.04	.05	.06	.07	.08	.09
2.1	.4821	.4826	.4830	.4834	.4838	.4842	.4846	.4850	.4854	.4857
2.2	.4861	.4864	.4868	.4871	.4875	.4878	.4881	.4884	.4887	.4890
2.3	.4893	.4896	.4898	.4901	.4904	.4906	.4909	.4911	.4913	.4916
2.4	.4918	.4920	.4922	.4925	.4927	.4929	.4931	.4932	.4934	.4936
2.5	.4938	.4940	.4941	.4943	.4945	.4946	.4948	.4949	.4951	.4952
2.6	.4953	.4955	.4956	.4957	.4959	.4960	.4961	.4962	.4963	.4964
2.7	.4965	.4966	.4967	.4968	.4969	.4970	.4971	.4972	.4973	.4974
2.8	.4974	.4975	.4976	.4977	.4977	.4978	.4979	.4979	.4980	.4981
2.9	.4981	.4982	.4982	.4983	.4984	.4984	.4985	.4985	.4986	.4986
3.0	.4987	.4987	.4987	.4988	.4988	.4989	.4989	.4989	.4990	.4990

Also, for $Z = 4.0$, 5.0, and 6.0, the areas are 0.49997, 0.4999997, and 0.499999999.

APPENDIX TABLE 8 Critical Values of t

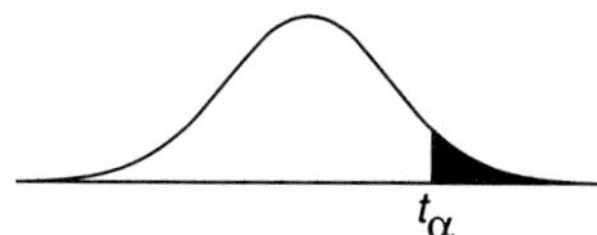

n	t_{100}	t_{050}	t_{025}	t_{010}	t_{005}	ν or d.f.
2	3.078	6.314	12.706	31.821	63.657	1
3	1.886	2.920	4.303	6.965	9.925	2
4	1.638	2.353	3.182	4.541	5.841	3
5	1.533	2.132	2.776	3.747	4.604	4
6	1.476	2.015	2.571	3.365	4.032	5
7	1.440	1.943	2.447	3.143	3.707	6
8	1.415	1.895	2.365	2.998	3.499	7
9	1.397	1.860	2.306	2.896	3.355	8
10	1.383	1.833	2.262	2.821	3.250	9
11	1.372	1.812	2.228	2.764	3.169	10
12	1.363	1.796	2.201	2.718	3.106	11
13	1.356	1.782	2.179	2.681	3.055	12
14	1.350	1.771	2.160	2.650	3.012	13
15	1.345	1.761	2.145	2.624	2.977	14
16	1.341	1.753	2.131	2.602	2.947	15
17	1.337	1.746	2.120	2.583	2.921	16
18	1.333	1.740	2.110	2.567	2.898	17
19	1.330	1.734	2.101	2.552	2.878	18
20	1.328	1.729	2.093	2.539	2.861	19
21	1.325	1.725	2.086	2.528	2.845	20
22	1.323	1.721	2.080	2.518	2.831	21
23	1.321	1.717	2.074	2.508	2.819	22
24	1.319	1.714	2.069	2.500	2.807	23
25	1.318	1.711	2.064	2.492	2.797	24
26	1.316	1.708	2.060	2.485	2.787	25
27	1.315	1.706	2.056	2.479	2.779	26
28	1.314	1.703	2.052	2.473	2.771	27
29	1.313	1.701	2.048	2.467	2.763	28
30	1.311	1.699	2.045	2.462	2.756	29
Inf.	1.282	1.645	1.960	2.326	2.576	Inf.

APPENDIX TABLE 9 Critical Values of Chi-Square

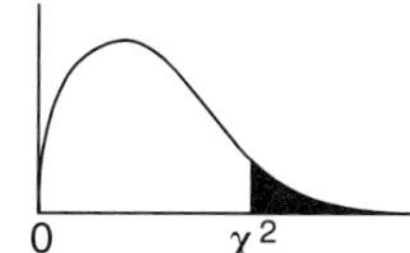

ν or d.f.	$\chi^2$0.995	$\chi^2$0.990	$\chi^2$0.975	$\chi^2$0.950	$\chi^2$0.900
1	0.0000393	0.0001571	0.0009821	0.0039321	0.0157908
2	0.0100251	0.0201007	0.0506356	0.102587	0.210720
3	0.0717212	0.114832	0.215795	0.351846	0.584375
4	0.206990	0.297110	0.484419	0.710721	1.063623
5	0.411740	0.554300	0.831211	1.145476	1.61031
6	0.675727	0.872085	1.237347	1.63539	2.20413
7	0.989265	1.239043	1.68987	2.16735	2.83311
8	1.344419	1.646482	2.17973	2.73264	3.48954
9	1.734926	2.087912	2.70039	3.32511	4.16816
10	2.15585	2.55821	3.24697	3.94030	4.86518
11	2.60321	3.05347	3.81575	4.57481	5.57779
12	3.07382	3.57056	4.40379	5.22603	6.30380
13	3.56503	4.10691	5.00874	5.89186	7.04150
14	4.07468	4.66043	5.62872	6.57063	7.78953
15	4.60094	5.22935	6.26214	7.26094	8.54675
16	5.14224	5.81221	6.90766	7.96164	9.31223
17	5.69724	6.40776	7.56418	8.67176	10.0852
18	6.26481	7.01491	8.23075	9.39046	10.8649
19	6.84398	7.63273	8.90655	10.1170	11.6509
20	7.43386	8.26040	9.59083	10.8508	12.4426
21	8.03366	8.89720	10.28293	11.5913	13.2396
22	8.64272	9.54249	10.9823	12.3380	14.0415
23	9.26042	10.19567	11.6885	13.0905	14.8479
24	9.88623	10.8564	12.4011	13.8484	15.6587
25	10.5197	11.5240	13.1197	14.6114	16.4734
26	11.1603	12.1981	13.8439	15.3791	17.2919
27	11.8076	12.8786	14.5733	16.1513	18.1138
28	12.4613	13.5648	15.3079	16.9279	18.9392
29	13.1211	14.2565	16.0471	17.7083	19.7677
30	13.7867	14.9535	16.7908	18.4926	20.5992
40	20.7065	22.1643	24.4331	26.5093	29.0505
50	27.9907	29.7067	32.3574	34.7642	37.6886
60	35.5346	37.4848	40.4817	43.1879	46.4589
70	43.2752	45.4418	48.7576	51.7393	55.3290
80	51.1720	53.5400	57.1532	60.3915	64.2778
90	59.1963	61.7541	65.6466	69.1260	73.2912
100	67.3276	70.0648	74.2219	77.9295	82.3581

APPENDIX TABLE 9 Critical Values of Chi-Square (continued)

$\chi^2 0.100$	$\chi^2 0.050$	$\chi^2 0.025$	$\chi^2 0.010$	$\chi^2 0.005$	ν or d.f.
2.70554	3.84146	5.02389	6.63490	7.87944	1
4.60517	5.99147	7.37776	9.21034	10.5966	2
6.25139	7.81473	9.34840	11.3449	12.8381	3
7.77944	9.48773	11.1433	13.2767	14.8602	4
9.23635	11.0705	12.8325	15.0863	16.7496	5
10.6446	12.5916	14.4494	16.8119	18.5476	6
12.0170	14.0671	16.0128	18.4753	20.2777	7
13.3616	15.5073	17.5346	20.0902	21.9550	8
14.6837	16.9190	19.0228	21.6660	23.5893	9
15.9871	18.3070	20.4831	23.2093	25.1882	10
17.2750	19.6751	21.9200	24.7250	26.7569	11
18.5494	21.0261	23.3367	26.2170	28.2995	12
19.8119	22.3621	24.7356	27.6883	29.8194	13
21.0642	23.6848	26.1190	29.1413	31.3193	14
22.3072	24.9958	27.4884	30.5779	32.8013	15
23.5418	26.2962	28.8454	31.9999	34.2672	16
24.7690	27.5871	30.1910	33.4087	35.7185	17
25.9894	28.8693	31.5264	34.8053	37.1564	18
27.2036	30.1435	32.8523	36.1908	38.5822	19
28.4120	31.4104	34.1696	37.5662	39.9968	20
29.6151	32.6705	35.4789	38.9321	41.4010	21
30.8133	33.9244	36.7807	40.2894	42.7956	22
32.0069	35.1725	38.0757	41.6384	44.1813	23
33.1963	36.4151	39.3641	42.9798	45.5585	24
34.3816	37.6525	40.6465	44.3141	46.9278	25
35.5631	38.8852	41.9232	45.6417	48.2899	26
36.7412	40.1133	43.1944	46.9630	49.6449	27
37.9159	41.3372	44.4607	48.2782	50.9933	28
39.0875	42.5569	45.7222	49.5879	52.3356	29
40.2560	43.7729	46.9792	50.8922	53.6720	30
51.8050	55.7585	59.3417	63.6907	66.7659	40
63.1671	67.5048	71.4202	76.1539	79.4900	50
74.3970	79.0819	83.2976	88.3794	91.9517	60
85.5271	90.5312	95.0231	100.425	104.215	70
96.5782	101.879	106.629	112.329	116.321	80
107.565	113.145	118.136	124.116	128.299	90
118.498	124.342	129.561	135.807	140.169	100

APPENDIX TABLE 10 5 and 1 Percent Values of the *F* Distribution

5% (Roman type) and 1% (boldface type)

d. f.	Degrees of Freedom (for greater mean square)																							d. f.
	1	2	3	4	5	6	7	8	9	10	11	12	14	16	20	24	30	40	50	75	100	200	500	∞	
1	161	200	216	225	230	234	237	239	241	242	243	244	245	246	248	249	250	251	252	253	253	254	254	254	1
	4,052	**4,999**	**5,403**	**5,625**	**5,764**	**5,859**	**5,928**	**5,981**	**6,022**	**6,056**	**6,082**	**6,106**	**6,142**	**6,169**	**6,208**	**6,234**	**6,261**	**6,286**	**6,302**	**6,323**	**6,334**	**6,352**	**6,361**	**6,366**	
2	18.51	19.00	19.16	19.25	19.30	19.33	19.36	19.37	19.38	19.39	19.40	19.41	19.42	19.43	19.44	19.45	19.46	19.47	19.47	19.48	19.49	19.49	19.50	19.50	2
	98.49	**99.00**	**99.17**	**99.25**	**99.30**	**99.33**	**99.36**	**99.37**	**99.39**	**99.40**	**99.41**	**99.42**	**99.43**	**99.44**	**99.45**	**99.46**	**99.47**	**99.48**	**99.48**	**99.49**	**99.49**	**99.49**	**99.49**	**99.50**	
3	10.13	9.55	9.28	9.12	9.01	8.94	8.88	8.84	8.81	8.78	8.76	8.74	8.71	8.69	8.66	8.64	8.62	8.60	8.58	8.57	8.56	8.54	8.54	8.53	3
	34.12	**30.82**	**29.46**	**28.71**	**28.24**	**27.91**	**27.67**	**27.49**	**27.34**	**27.23**	**27.13**	**27.05**	**26.92**	**26.83**	**26.69**	**26.60**	**26.50**	**26.41**	**26.35**	**26.27**	**26.23**	**26.18**	**26.14**	**26.12**	
4	7.71	6.94	6.59	6.39	6.26	6.16	6.09	6.04	6.00	5.96	5.93	5.91	5.87	5.84	5.80	5.77	5.74	5.71	5.70	5.68	5.66	5.65	5.64	5.63	4
	21.20	**18.00**	**16.69**	**15.98**	**15.52**	**15.21**	**14.98**	**14.80**	**14.66**	**14.54**	**14.45**	**14.37**	**14.24**	**14.15**	**14.02**	**13.93**	**13.83**	**13.74**	**13.69**	**13.61**	**13.57**	**13.52**	**13.48**	**13.46**	
5	6.61	5.79	5.41	5.19	5.05	4.95	4.88	4.82	4.78	4.74	4.70	4.68	4.64	4.60	4.56	4.53	4.50	4.46	4.44	4.42	4.40	4.38	4.37	4.36	5
	16.26	**13.27**	**12.06**	**11.39**	**10.97**	**10.67**	**10.45**	**10.29**	**10.15**	**10.05**	**9.96**	**9.89**	**9.77**	**9.68**	**9.55**	**9.47**	**9.38**	**9.29**	**9.24**	**9.17**	**9.13**	**9.07**	**9.04**	**9.02**	
6	5.99	5.14	4.76	4.53	4.39	4.28	4.21	4.15	4.10	4.06	4.03	4.00	3.96	3.92	3.87	3.84	3.81	3.77	3.75	3.72	3.71	3.69	3.68	3.67	6
	13.74	**10.92**	**9.78**	**9.15**	**8.75**	**8.47**	**8.26**	**8.10**	**7.98**	**7.87**	**7.79**	**7.72**	**7.60**	**7.52**	**7.39**	**7.31**	**7.23**	**7.14**	**7.09**	**7.02**	**6.99**	**6.94**	**6.90**	**6.88**	
7	5.59	4.74	4.34	4.12	3.97	3.87	3.79	3.73	3.68	3.63	3.60	3.57	3.52	3.49	3.44	3.41	3.38	3.34	3.32	3.29	3.28	3.25	3.24	3.23	7
	12.25	**9.55**	**8.45**	**7.85**	**7.46**	**7.19**	**7.00**	**6.84**	**6.71**	**6.62**	**6.54**	**6.47**	**6.35**	**6.27**	**6.15**	**6.07**	**5.98**	**5.90**	**5.85**	**5.78**	**5.75**	**5.70**	**5.67**	**5.65**	
8	5.32	4.46	4.07	3.84	3.69	3.58	3.50	3.44	3.39	3.34	3.31	3.28	3.23	3.20	3.15	3.12	3.08	3.05	3.03	3.00	2.98	2.96	2.94	2.93	8
	11.26	**8.65**	**7.59**	**7.01**	**6.63**	**6.37**	**6.19**	**6.03**	**5.91**	**5.82**	**5.74**	**5.67**	**5.56**	**5.48**	**5.36**	**5.28**	**5.20**	**5.11**	**5.06**	**5.00**	**4.96**	**4.91**	**4.88**	**4.86**	
9	5.12	4.26	3.86	3.63	3.48	3.37	3.29	3.23	3.18	3.13	3.10	3.07	3.02	2.98	2.93	2.90	2.86	2.82	2.80	2.77	2.76	2.73	2.72	2.71	9
	10.56	**8.02**	**6.99**	**6.42**	**6.06**	**5.80**	**5.62**	**5.47**	**5.35**	**5.26**	**5.18**	**5.11**	**5.00**	**4.92**	**4.80**	**4.73**	**4.64**	**4.56**	**4.51**	**4.45**	**4.41**	**4.36**	**4.33**	**4.31**	
10	4.96	4.10	3.71	3.48	3.33	3.22	3.14	3.07	3.02	2.97	2.94	2.91	2.86	2.82	2.77	2.74	2.70	2.67	2.64	2.61	2.59	2.56	2.55	2.54	10
	10.04	**7.56**	**6.55**	**5.99**	**5.64**	**5.39**	**5.21**	**5.06**	**4.95**	**4.85**	**4.78**	**4.71**	**4.60**	**4.52**	**4.41**	**4.33**	**4.25**	**4.17**	**4.12**	**4.05**	**4.01**	**3.96**	**3.93**	**3.91**	
11	4.84	3.98	3.59	3.36	3.20	3.09	3.01	2.95	2.90	2.86	2.82	2.79	2.74	2.70	2.65	2.61	2.57	2.53	2.50	2.47	2.45	2.42	2.41	2.40	11
	9.65	**7.20**	**6.22**	**5.67**	**5.32**	**5.07**	**4.88**	**4.74**	**4.63**	**4.54**	**4.46**	**4.40**	**4.29**	**4.21**	**4.10**	**4.02**	**3.94**	**3.86**	**3.80**	**3.74**	**3.70**	**3.66**	**3.62**	**3.60**	
12	4.75	3.88	3.49	3.26	3.11	3.00	2.92	2.85	2.80	2.76	2.72	2.69	2.64	2.60	2.54	2.50	2.46	2.42	2.40	2.36	2.35	2.32	2.31	2.30	12
	9.33	**6.93**	**5.95**	**5.41**	**5.06**	**4.82**	**4.65**	**4.50**	**4.39**	**4.30**	**4.22**	**4.16**	**4.05**	**3.98**	**3.86**	**3.78**	**3.70**	**3.61**	**3.56**	**3.49**	**3.46**	**3.41**	**3.38**	**3.36**	

APPENDIX TABLE 10 5 and 1 Percent Values of the F Distribution (continued)

5% (Roman type) and 1% (boldface type)

	Degrees of Freedom (for greater mean square)																								
d. f.	1	2	3	4	5	6	7	8	9	10	11	12	14	16	20	24	30	40	50	75	100	200	500	∞	d. f.
13	4.67	3.80	3.41	3.18	3.02	2.92	2.0	2.77	2.72	2.67	2.63	2.60	2.55	2.51	2.46	2.42	238	2.34	2.32	2.28	2.26	2.24	2.22	2.21	13
	9.07	**6.70**	**5.74**	**5.20**	**4.86**	**4.62**	**4.44**	**4.30**	**4.19**	**4.10**	**4.02**	**3.96**	**3.85**	**3.78**	**3.67**	**3.59**	**3.51**	**3.42**	**3.37**	**3.30**	**3.27**	**3.21**	**3.18**	**3.16**	
14	4.60	3.74	3.34	3.11	2.96	2.85	2.77	2.70	2.65	2.60	2.56	2.53	2.48	2.44	2.39	2.35	2.31	2.27	2.24	2.21	2.19	2.16	2.14	2.13	14
	8.86	**6.51**	**5.56**	**5.03**	**4.69**	**4.46**	**4.28**	**4.14**	**4.03**	**3.94**	**3.86**	**3.80**	**3.70**	**3.62**	**3.51**	**3.43**	**3.34**	**3.26**	**3.21**	**3.14**	**3.11**	**3.06**	**3.02**	**3.00**	
15	4.54	3.68	3.29	3.06	2.90	2.79	2.70	2.64	2.59	2.55	2.51	2.48	2.43	2.39	2.33	2.29	2.25	2.21	2.18	2.15	2.12	2.10	2.08	2.07	15
	8.68	**6.36**	**5.42**	**4.89**	**4.56**	**4.32**	**4.14**	**4.00**	**3.89**	**3.80**	**3.73**	**3.67**	**3.56**	**3.48**	**3.36**	**3.29**	**3.20**	**3.12**	**3.07**	**3.00**	**2.97**	**2.92**	**2.89**	**2.87**	
16	4.49	3.63	3.24	3.01	2.85	2.74	2.66	2.59	2.54	2.49	2.45	2.42	2.37	2.33	2.28	2.24	2.20	2.16	2.13	2.09	2.07	2.04	2.02	2.01	16
	8.53	**6.23**	**5.29**	**4.77**	**4.44**	**4.20**	**4.03**	**3.89**	**3.78**	**3.69**	**3.61**	**3.55**	**3.45**	**3.37**	**3.25**	**3.18**	**3.10**	**3.01**	**2.96**	**2.98**	**2.86**	**2.80**	**2.77**	**2.75**	
17	4.45	3.59	3.20	2.96	2.81	2.70	2.62	2.55	2.50	2.45	2.41	2.38	2.33	2.29	2.23	2.19	2.15	2.11	2.08	2.04	2.02	1.99	1.97	1.96	17
	8.40	**6.11**	**5.18**	**4.67**	**4.34**	**4.10**	**3.93**	**3.79**	**3.68**	**3.59**	**3.52**	**3.45**	**3.35**	**3.27**	**3.16**	**3.08**	**3.00**	**2.92**	**2.86**	**2.79**	**2.76**	**2.70**	**2.67**	**2.65**	
18	4.41	3.55	3.16	2.93	2.77	2.66	2.58	2.51	2.46	2.41	2.37	2.34	2.29	2.25	2.19	2.15	2.11	2.07	2.04	2.00	1.98	1.95	1.93	1.92	18
	8.28	**6.01**	**5.09**	**4.58**	**4.25**	**4.01**	**3.85**	**3.71**	**3.60**	**3.51**	**3.44**	**3.37**	**3.27**	**3.19**	**3.07**	**3.00**	**2.91**	**2.83**	**2.78**	**2.71**	**2.68**	**2.62**	**2.59**	**2.57**	
19	4.38	3.52	3.13	2.90	2.74	2.63	2.55	2.48	2.43	2.38	2.34	2.31	2.26	2.21	2.15	2.11	2.07	2.02	2.00	1.96	1.94	1.91	1.90	1.88	19
	8.18	**5.93**	**5.01**	**4.50**	**4.17**	**3.94**	**3.77**	**3.63**	**3.52**	**3.43**	**3.36**	**3.30**	**3.19**	**3.12**	**3.00**	**2.92**	**2.84**	**2.76**	**2.70**	**2.63**	**2.60**	**2.54**	**2.51**	**2.49**	
20	4.35	3.49	3.10	2.87	2.71	2.60	2.52	2.45	2.40	2.35	2.31	2.28	2.23	2.18	2.12	2.08	2.04	1.99	1.96	1.92	1.90	1.87	1.85	1.84	20
	8.10	**5.85**	**4.94**	**4.43**	**4.10**	**3.87**	**3.71**	**3.56**	**3.45**	**3.37**	**3.30**	**3.23**	**3.13**	**3.05**	**2.94**	**2.86**	**2.77**	**2.69**	**2.63**	**2.56**	**2.53**	**2.47**	**2.44**	**2.42**	
21	4.32	3.47	3.07	2.84	2.68	2.57	2.49	2.42	2.37	2.32	2.28	2.25	2.20	2.15	2.09	2.05	2.00	1.96	1.93	1.89	1.87	1.84	1.82	1.81	21
	8.02	**5.78**	**4.87**	**4.37**	**4.04**	**3.81**	**3.65**	**3.51**	**3.40**	**3.31**	**3.24**	**3.17**	**3.07**	**2.99**	**2.88**	**2.80**	**2.72**	**2.63**	**2.58**	**2.51**	**2.47**	**2.42**	**2.38**	**2.36**	
22	4.30	3.44	3.05	2.82	2.66	2.55	2.47	2.40	2.35	2.30	2.26	2.23	2.18	2.13	2.07	2.03	1.98	1.93	1.91	1.87	1.84	1.81	1.80	1.78	22
	7.94	**5.72**	**4.82**	**4.31**	**3.99**	**3.76**	**3.59**	**3.45**	**3.35**	**3.26**	**3.18**	**3.12**	**3.02**	**2.94**	**2.83**	**2.75**	**2.67**	**2.58**	**2.53**	**2.46**	**2.42**	**2.37**	**2.33**	**2.31**	
23	4.28	3.42	3.03	2.80	2.64	2.53	2.45	2.38	2.32	2.28	2.24	2.20	2.14	2.10	2.04	2.00	1.96	1.91	1.88	1.84	1.82	1.79	1.77	1.76	23
	7.88	**5.66**	**4.76**	**4.26**	**3.94**	**3.71**	**3.54**	**3.41**	**3.30**	**3.21**	**3.14**	**3.07**	**2.97**	**2.89**	**2.78**	**2.70**	**2.62**	**2.53**	**2.48**	**2.41**	**2.37**	**2.32**	**2.28**	**2.26**	
24	4.26	3.40	3.01	2.78	2.62	2.51	2.43	2.36	2.30	2.26	2.22	2.18	2.13	2.09	2.02	1.98	1.94	1.89	1.86	1.82	1.80	1.76	1.74	1.73	24
	7.82	**5.61**	**4.72**	**4.22**	**3.90**	**3.67**	**3.50**	**3.36**	**3.25**	**3.17**	**3.09**	**3.03**	**2.93**	**2.85**	**2.74**	**2.66**	**2.58**	**2.49**	**2.44**	**2.36**	**2.33**	**2.27**	**2.23**	**2.21**	

APPENDIX TABLE 10 5 and 1 Percent Values of the *F* Distribution (continued)

5% (Roman type) and 1% (boldface type)

d. f.	Degrees of Freedom (for greater mean square)																								d. f.
	1	2	3	4	5	6	7	8	9	10	11	12	14	16	20	24	30	40	50	75	100	200	500	∞	
25	4.24	3.38	2.99	2.76	2.60	2.49	2.41	2.34	2.28	2.24	2.20	2.16	2.11	2.06	2.00	1.96	1.92	1.87	1.84	1.80	1.77	1.74	1.72	1.71	25
	7.77	**5.57**	**4.68**	**4.18**	**3.86**	**3.63**	**3.46**	**3.32**	**3.21**	**3.13**	**3.05**	**2.99**	**2.89**	**2.81**	**2.70**	**2.62**	**2.54**	**2.45**	**2.40**	**2.32**	**2.29**	**2.23**	**2.19**	**2.17**	
26	4.22	3.37	2.98	2.74	2.59	2.47	2.39	2.32	2.27	2.22	2.18	2.15	2.10	2.05	1.99	1.95	1.90	1.85	1.82	1.78	1.76	1.72	1.70	1.69	26
	7.72	**5.53**	**4.64**	**4.14**	**3.82**	**3.59**	**3.42**	**3.29**	**3.17**	**3.09**	**3.02**	**2.96**	**2.86**	**2.77**	**2.66**	**2.58**	**2.50**	**2.41**	**2.36**	**2.28**	**2.25**	**2.19**	**2.15**	**2.13**	
27	4.21	3.35	2.96	2.73	2.57	2.46	2.37	2.30	2.25	2.20	2.16	2.13	2.08	2.03	1.97	1.93	1.88	1.84	1.80	1.76	1.74	1.71	1.68	1.67	27
	7.68	**5.49**	**4.60**	**4.11**	**3.79**	**3.56**	**3.39**	**3.26**	**3.14**	**3.06**	**2.98**	**2.93**	**2.83**	**2.74**	**2.63**	**2.55**	**2.47**	**2.38**	**2.33**	**2.25**	**2.21**	**2.16**	**2.12**	**2.10**	
28	4.20	3.34	2.95	2.71	2.56	2.44	2.36	2.29	2.24	2.19	2.15	2.12	2.06	2.02	1.96	1.91	1.87	1.81	1.78	1.75	1.72	1.69	1.67	1.65	28
	7.64	**5.45**	**4.57**	**4.07**	**3.76**	**3.53**	**3.36**	**3.23**	**3.11**	**3.03**	**2.95**	**2.90**	**2.80**	**2.71**	**2.60**	**2.52**	**2.44**	**2.35**	**2.30**	**2.22**	**2.18**	**2.13**	**2.09**	**2.06**	
29	4.18	3.33	2.93	2.70	2.54	2.43	2.35	2.28	2.22	2.18	2.14	2.10	2.05	2.00	1.94	1.90	1.85	1.80	1.77	1.73	1.71	1.68	1.65	1.64	29
	7.60	**5.42**	**4.54**	**4.04**	**3.73**	**3.50**	**3.33**	**3.20**	**3.08**	**3.00**	**2.92**	**2.87**	**2.77**	**2.68**	**2.57**	**2.49**	**2.41**	**2.32**	**2.27**	**2.19**	**2.15**	**2.10**	**2.06**	**2.03**	
30	4.17	3.32	2.92	2.69	2.53	2.42	2.34	2.27	2.21	2.16	2.12	2.09	2.04	1.99	1.93	1.89	1.84	1.79	1.76	1.72	1.69	1.66	1.64	1.62	30
	7.56	**5.39**	**4.51**	**4.02**	**3.70**	**3.47**	**3.30**	**3.17**	**3.06**	**2.98**	**2.90**	**2.84**	**2.74**	**2.66**	**2.55**	**2.47**	**2.38**	**2.29**	**2.24**	**2.16**	**2.13**	**2.07**	**2.03**	**2.01**	
32	4.15	3.30	2.90	2.67	2.51	2.40	2.32	2.25	2.19	2.14	2.10	2.07	2.02	1.97	1.91	1.86	1.82	1.76	1.74	1.69	1.67	1.64	1.61	1,59	32
	7.50	**5.34**	**4.46**	**3.97**	**3.66**	**3.42**	**3.25**	**3.12**	**3.01**	**2.94**	**2.86**	**2.80**	**2.70**	**2.62**	**2.51**	**2.42**	**2.34**	**2.25**	**2.20**	**2.12**	**2.08**	**2.02**	**1.98**	**1.96**	
34	4.13	3.28	2.88	2.65	2.49	2.38	2.30	2.23	2.17	2.12	2.08	2.05	2.00	1.95	1.89	1.84	1.80	1.74	1.71	1.67	1.64	1.61	1.59	1.57	34
	7.44	**5.29**	**4.42**	**3.93**	**3.61**	**3.38**	**3.21**	**3.08**	**2.97**	**2.89**	**2.82**	**2.76**	**2.66**	**2.58**	**2.47**	**2.38**	**2.30**	**2.21**	**2.15**	**2.08**	**2.04**	**1.98**	**1.94**	**1.91**	
36	4.11	3.26	2.86	2.63	2.48	2.36	2.28	2.21	2.15	2.10	2.06	2.03	1.98	1.93	1.87	1.82	1.78	1.72	1.69	1.65	1.62	1.59	1.56	1.55	36
	7.39	**5.25**	**4.38**	**3.89**	**3.58**	**3.35**	**3.18**	**3.04**	**2.94**	**2.86**	**2.78**	**2.72**	**2.62**	**2.54**	**2.43**	**2.35**	**2.26**	**2.17**	**2.12**	**2.04**	**2.00**	**1.94**	**1.90**	**1.87**	
38	4.10	3.25	2.85	2.62	2.46	2.35	2.26	2.19	2.14	2.09	2.05	2.02	1.96	1.92	1.85	1.80	1.76	1.71	1.67	1.63	1.60	1.57	1.54	1.53	38
	7.35	**5.21**	**4.34**	**3.86**	**3.54**	**3.32**	**3.15**	**3.02**	**2.91**	**2.82**	**2.75**	**2.69**	**2.59**	**2.51**	**2.40**	**2.32**	**2.22**	**2.14**	**2.08**	**2.00**	**1.97**	**1.90**	**1.86**	**1.84**	
40	4.07	3.23	2.84	2.61	2.45	2.34	2.25	2.18	2.12	2.07	2.04	2.00	1.95	1.90	1.84	1.79	1.74	1.69	1.66	1.61	1.59	1.55	1.53	1.51	40
	7.31	**5.18**	**4.31**	**3.83**	**3.51**	**3.29**	**3.12**	**2.99**	**2.88**	**2.80**	**2.73**	**2.66**	**2.56**	**2.49**	**2.37**	**2.29**	**2.20**	**2.11**	**2.05**	**1.97**	**1.94**	**1.88**	**1.84**	**1.81**	
42	4.07	3.22	2.83	2.59	2.44	2.32	2.24	2.17	2.11	2.06	2.02	1.99	1.94	1.89	1.82	1.78	1.73	1.68	1.64	1.60	1.57	1.54	1.51	1.49	42
	7.27	**5.15**	**4.29**	**3.80**	**3.49**	**3.26**	**3.10**	**2.96**	**2.86**	**2.77**	**2.70**	**2.64**	**2.54**	**2.46**	**2.35**	**2.26**	**2.17**	**2.08**	**2.02**	**1.94**	**1.91**	**1.85**	**1.80**	**1.78**	

APPENDIX TABLE 10 5 and 1 Percent Values of the *F* Distribution (continued)

5% (Roman type) and 1% (boldface type)

	Degrees of Freedom (for greater mean square)																								
d. f.	1	2	3	4	5	6	7	8	9	10	11	12	14	16	20	24	30	40	50	75	100	200	500	∞	d. f.
44	4.06	3.21	2.82	2.58	2.43	2.31	2.23	2.16	2.10	2.05	2.01	1.98	1.92	1.88	1.81	1.76	1.72	1.66	1.63	1.58	1.56	1.52	1.50	1.48	44
	7.24	**5.12**	**4.26**	**3.78**	**3.46**	**3.24**	**3.07**	**2.94**	**2.84**	**2.75**	**2.68**	**2.62**	**2.52**	**2.44**	**2.32**	**2.24**	**2.15**	**2.06**	**2.00**	**1.92**	**1.88**	**1.82**	**1.78**	**1.75**	
46	4.05	3.20	2.81	2.57	2.42	2.30	2.22	2.14	2.09	2.04	2.00	1.97	1.91	1.87	1.80	1.75	1.71	1.65	1.62	1.57	1.54	1.51	1.48	1.46	46
	7.21	**5.10**	**4.24**	**3.76**	**3.44**	**3.22**	**3.05**	**2.92**	**2.82**	**2.73**	**2.66**	**2.60**	**2.50**	**2.42**	**2.30**	**2.22**	**2.13**	**2.04**	**1.98**	**1.90**	**1.86**	**1.80**	**1.76**	**1.72**	
48	4.04	3.19	2.80	2.56	2.41	2.30	2.21	2.14	2.08	2.03	1.99	1.96	1.90	1.86	1.79	1.74	1.70	1.64	1.61	1.56	1.53	1.50	1.47	1.45	48
	7.19	**5.08**	**4.22**	**3.74**	**3.42**	**3.20**	**3.04**	**2.90**	**2.80**	**2.71**	**2.64**	**2.58**	**2.48**	**2.40**	**2.28**	**2.20**	**2.11**	**2.02**	**1.96**	**1.88**	**1.84**	**1.78**	**1.73**	**1.70**	
50	4.03	3.18	2.79	2.56	2.40	2.29	2.20	2.13	2.07	2.02	1.98	1.95	1.90	1.85	1.78	1.74	1.69	1.63	1.60	1.55	1.52	1.48	1.46	1.44	50
	7.17	**5.06**	**4.20**	**3.72**	**3.41**	**3.18**	**3.02**	**2.88**	**2.78**	**2.70**	**2.62**	**2.56**	**2.46**	**2.39**	**2.26**	**2.18**	**2.10**	**2.00**	**1.94**	**1.86**	**1.82**	**1.76**	**1.71**	**1.68**	
55	4.02	3.17	2.78	2.54	2.38	2.27	2.18	2.11	2.05	2.00	1.97	1.93	1.88	1.83	1.76	1.72	1.67	1.61	1.58	1.52	1.50	1.46	1.43	1.41	55
	7.12	**5.01**	**4.16**	**3.68**	**3.37**	**3.15**	**2.98**	**2.85**	**2.75**	**2.66**	**2.59**	**2.53**	**2.43**	**2.35**	**2.23**	**2.15**	**2.06**	**1.96**	**1.90**	**1.82**	**1.78**	**1.71**	**1.66**	**1.64**	
60	4.00	3.15	2.76	2.52	2.37	2.25	2.17	2.10	2.04	1.99	1.95	1.92	1.86	1.81	1.75	1.70	1.65	1.59	1.56	1.50	1.48	1.44	1.41	1.39	60
	7.08	**4.98**	**4.13**	**3.65**	**3.34**	**3.12**	**2.95**	**2.82**	**2.72**	**2.63**	**2.56**	**2.50**	**2.40**	**2.32**	**2.20**	**2.12**	**2.03**	**1.93**	**1.87**	**1.79**	**1.74**	**1.68**	**1.63**	**1.60**	
65	3.99	3.14	2.75	2.51	2.36	2.24	2.15	2.08	2.02	1.98	1.94	1.90	1.85	1.80	1.73	1.68	1.63	1.57	1.54	1.49	1.46	1.42	1.39	1.37	65
	7.04	**4.95**	**4.10**	**3.62**	**3.31**	**3.09**	**2.93**	**2.79**	**2.70**	**2.61**	**2.54**	**2.47**	**2.37**	**2.30**	**2.18**	**2.09**	**2.00**	**1.90**	**1.84**	**1.76**	**1.71**	**1.64**	**1.60**	**1.56**	
70	3.98	3.13	2.74	2.50	2.35	2.23	2.14	2.07	2.01	1.97	1.93	1.89	1.84	1.79	1.72	1.67	1.62	1.56	1.53	1.47	1.45	1.40	1.37	1.35	70
	7.01	**4.92**	**4.08**	**3.60**	**3.29**	**3.07**	**2.91**	**2.77**	**2.67**	**2.59**	**2.51**	**2.45**	**2.35**	**2.28**	**2.15**	**2.07**	**1.98**	**1.88**	**1.82**	**1.74**	**1.69**	**1.62**	**1.56**	**1.53**	
80	3.96	3.11	2.72	2.48	2.33	2.21	2.12	2.05	1.99	1.95	1.91	1.88	1.82	1.77	1.70	1.65	1.60	1.54	1.51	1.45	1.42	1.38	1.35	1.32	80
	6.96	**4.88**	**4.04**	**3.56**	**3.25**	**3.04**	**2.87**	**2.74**	**2.64**	**2.55**	**2.48**	**2.41**	**2.32**	**2.24**	**2.11**	**2.03**	**1.94**	**1.84**	**1.78**	**1.70**	**1.65**	**1.57**	**1.52**	**1.49**	
100	3.94	3.09	2.70	2.46	2.30	2.19	2.10	2.03	1.97	1.92	1.88	1.85	1.79	1.75	1.68	1.63	1.57	1.51	1.48	1.42	1.39	1.34	1.30	1.28	100
	6.90	**4.82**	**3.98**	**3.51**	**3.20**	**2.99**	**2.82**	**2.69**	**2.59**	**2.51**	**2.43**	**2.36**	**2.26**	**2.19**	**2.06**	**1.98**	**1.89**	**1.79**	**1.73**	**1.64**	**1.59**	**1.51**	**1.46**	**1.43**	
125	3.92	3.07	2.68	2.44	2.29	2.17	2.08	2.01	1.95	1.90	1.86	1.83	1.77	1.72	1.65	1.60	1.55	1.49	1.45	1.39	1.36	1.31	1.27	1.25	125
	6.84	**4.78**	**3.94**	**3.47**	**3.17**	**2.95**	**2.79**	**2.65**	**2.56**	**2.47**	**2.40**	**2.33**	**2.23**	**2.15**	**2.03**	**1.94**	**1.85**	**1.75**	**1.68**	**1.59**	**1.54**	**1.46**	**1.40**	**1.37**	

APPENDIX TABLE 10 5 and 1 Percent Values of the *F* Distribution (continued)

5% (Roman type) and 1% (boldface type)

	Degrees of Freedom (for greater mean square)																								
d. f.	1	2	3	4	5	6	7	8	9	10	11	12	14	16	20	24	30	40	50	75	100	200	500	∞	d. f.
150	3.91	3.06	2.67	2.43	2.27	2.16	2.07	2.00	1.94	1.89	1.85	1.82	1.76	1.71	1.64	1.59	1.54	1.47	1.44	1.37	1.34	1.29	1.25	1.22	150
	6.81	**4.75**	**3.91**	**3.44**	**3.14**	**2.92**	**2.76**	**2.62**	**2.53**	**2.44**	**2.37**	**2.30**	**2.20**	**2.12**	**2.00**	**1.91**	**1.83**	**1.72**	**1.66**	**1.56**	**1.51**	**1.43**	**1.37**	**1.33**	
200	3.89	3.04	2.65	2.41	2.26	2.14	2.05	1.98	1.92	1.87	1.83	1.80	1.74	1.69	1.62	1.57	1.52	1.45	1.42	1.35	1.32	1.26	1.22	1.19	200
	6.76	**4.71**	**3.88**	**3.41**	**3.11**	**2.90**	**2.73**	**2.60**	**2.50**	**2.41**	**2.34**	**2.28**	**2.17**	**2.09**	**1.97**	**1.88**	**1.79**	**1.69**	**1.62**	**1.53**	**1.48**	**1.39**	**1.33**	**1.28**	
400	3.86	3.02	2.62	2.39	2.23	2.12	2.03	1.96	1.90	1.85	1.81	1.78	1.72	1.67	1.60	1.54	1.49	1.42	1.38	1.32	1.28	1.22	1.16	1.13	400
	6.70	**4.66**	**3.83**	**3.36**	**3.06**	**2.85**	**2.69**	**2.55**	**2.46**	**2.37**	**2.29**	**2.23**	**2.12**	**2.04**	**1.92**	**1.84**	**1.74**	**1.64**	**1.57**	**1.47**	**1.42**	**1.32**	**1.24**	**1.19**	
1000	3.85	3.00	2.61	2.38	2.22	2.10	2.02	1.95	1.89	1.84	1.80	1.76	1.70	1.65	1.58	1.53	1.47	1.41	1.36	1.30	1.26	1.19	1.13	1.08	1000
	6.66	**4.62**	**3.80**	**3.34**	**3.04**	**2.82**	**2.66**	**2.53**	**2.43**	**2.34**	**2.26**	**2.20**	**2.09**	**2.01**	**1.89**	**1.81**	**1.71**	**1.61**	**1.54**	**1.44**	**1.38**	**1.28**	**1.19**	**1.11**	
∞	3.84	2.99	2.60	2.37	2.21	2.09	2.01	1.94	1.88	1.83	1.79	1.75	1.69	1.64	1.57	1.52	1.46	1.40	1.35	1.28	1.24	1.17	1.11	1.00	∞
	6.64	**4.60**	**3.78**	**3.32**	**3.02**	**2.80**	**2.64**	**2.51**	**2.41**	**2.32**	**2.24**	**2.18**	**2.07**	**1.99**	**1.87**	**1.79**	**1.69**	**1.59**	**1.52**	**1.41**	**1.36**	**1.25**	**1.15**	**1.00**	

The function $F = e$, with exponent $2z$, is computed in part from Fisher's table VI (7). Additional entries are by interpolation, mostly graphical.

APPENDIX TABLE 11 Selected Latin Squares

3 × 3	4 × 4: 1	4 × 4: 2	4 × 4: 3	4 × 4: 4
A B C	*A B C D*	*A B C D*	*A B C D*	*A B C D*
C A B	*D A B C*	*B A D C*	*B D A C*	*B A D C*
B C A	*C D A B*	*D C A B*	*C A D B*	*C D A B*
	B C D A	*C D B A*	*D C B A*	*D C B A*

5 × 5	6 × 6	7 × 7
A B C D E	*A B C D E F*	*A B C D E F G*
E A B C D	*F A B C D E*	*G A B C D E F*
C D E A B	*E F A B C D*	*F G A B C D E*
D E A B C	*D E F A B C*	*E F G A B C D*
B C D E A	*C D E F A B*	*D E F G A B C*
	B C D E F A	*C D E F G A B*
		B C D E F G A

8 × 8	9 × 9
A B C D E F G H	*A B C D E F G H I*
H A B C D E F G	*I A B C D E F G H*
G H A B C D E F	*H I A B C D E F G*
F G H A B C D E	*G H I A B C D E F*
E F G H A B C D	*F G H I A B C D E*
D E F G H A B C	*E F G H I A B C D*
C D E F G H A B	*D E F G H I A B C*
B C D E F G H A	*C D E F G H I A B*
	B C D E F G H I A

APPENDIX TABLE 12 Critical Values of *r* in the Runs Test

The body of the table contains critical values of r at the 0.05 level of significance for values of n_1 and n_2.

$n_1 \mid n_2$	2	3	4	5	6	7	8	9	10	11	12	13	14	15	16	17	18	19	20
2											2	2	2	2	2	2	2	2	2
3					2	2	2	2	2	2	2	2	2	3	3	3	3	3	3
4				2	2	2	3	3	3	3	3	3	3	3	4	4	4	4	4
5			2	2	3	3	3	3	3	4	4	4	4	4	4	4	5	5	5
6		2	2	3	3	3	3	4	4	4	4	5	5	5	5	5	5	6	6
7		2	2	3	3	3	4	4	5	5	5	5	5	6	6	6	6	6	6
8		2	3	3	3	4	4	5	5	5	6	6	6	6	6	7	7	7	7
9		2	3	3	4	4	5	5	5	6	6	6	7	7	7	7	8	8	8
10		2	3	3	4	5	5	5	6	6	7	7	7	7	8	8	8	8	9
11		2	3	4	4	5	5	6	6	7	7	7	8	8	8	9	9	9	9
12	2	2	3	4	4	5	6	6	7	7	7	8	8	8	9	9	9	10	10
13	2	2	3	4	5	5	6	6	7	7	8	8	9	9	9	10	10	10	10
14	2	2	3	4	5	5	6	7	7	8	8	9	9	9	10	10	10	11	11
15	2	3	3	4	5	6	6	7	7	8	8	9	9	10	10	11	11	11	12
16	2	3	4	4	5	6	6	7	8	8	9	9	10	10	11	11	11	12	12
17	2	3	4	4	5	6	7	7	8	9	9	10	10	11	11	11	12	12	13
18	2	3	4	5	5	6	7	8	8	9	9	10	10	11	11	12	12	13	13
19	2	3	4	5	6	6	7	8	8	9	10	10	11	11	12	12	13	13	13
20	2	3	4	5	6	6	7	8	9	9	10	10	11	12	12	13	13	13	14

APPENDIX TABLE 13 Critical Values of Spearman's Rank Correlation Coefficient

n	$\alpha = 0.10$	$\alpha = 0.05$	$\alpha = 0.02$	$\alpha = 0.01$
5	0.900			
6	0.829	0.886	0.943	
7	0.714	0.786	0.893	
8	0.643	0.738	0.833	0.881
9	0.600	0.683	0.783	0.833
10	0.564	0.648	0.745	0.794
11	0.523	0.623	0.736	0.818
12	0.497	0.591	0.703	0.780
13	0.475	0.566	0.673	0.745
14	0.457	0.545	0.646	0.716
15	0.441	0.525	0.623	0.689
16	0.425	0.507	0.601	0.666
17	0.412	0.490	0.582	0.645
18	0.399	0.476	0.564	0.625
19	0.388	0.462	0.549	0.608
20	0.377	0.450	0.534	0.591
21	0.368	0.438	0.521	0.576
22	0.359	0.428	0.508	0.562
23	0.351	0.418	0.496	0.549
24	0.343	0.409	0.485	0.537
25	0.336	0.400	0.475	0.526
26	0.329	0.392	0.465	0.515
27	0.323	0.385	0.456	0.505
28	0.317	0.377	0.448	0.496
29	0.311	0.370	0.440	0.487
30	0.305	0.364	0.432	0.478

Statistical Tables Endnotes

1. From *Statistical Analysis for Administrative Decisions, 3rd edition,* by Clark & Schkade © 1978. Reprinted with permission of South-Western Educational Publishing, a division of Thomson Learning. FAX 800 730-2215. *Table 1: Random Numbers*
2. From *Statistical Analysis for Administrative Decisions, 3rd edition,* by Clark & Schkade © 1978. Reprinted with permission of South-Western Education Publishing, a Division of Thomson Learning. FAX 800 730-2215. *Table 6: Poisson Distribution*
3. From *Biometrika, Vol 32, p. 300, 1941, Table of Percentage Points of the t-distribution,* by M. Merrington. Reprinted with permission of Oxford University Press, Oxford, UK. *Table 8: Critical Values of* t
4. From *Statistical Analysis for Administrative Decisions, 3rd edition,* by Clark & Schkade © 1978. Reprinted with permission of South-Western Educational Publishing, a Division of Thomson Learning. FAX 800 730-2215. *Table 10: 5 and 1 Percent Values for the* f *Distribution*
5. From *Annals of Mathematical Statistics, Vol 14, pp. 83–86, 1943, Tables for Testing Randomness of Grouping in a Sequence of Alternatives,* by Swed & Eisenhart. Reprinted with permission of Institute of Mathematical Statistics. FAX 510 783-4131. *Table 12: Critical Values of* r *in the Runs Test.*
6. From *Statistics for Management and Economics, 1st edition,* by MendenhallReinmuth © 1971. Reprinted with permission of Brooks/Cole Publishing, a division of Thomson Learning. FAX 800 730-2215. *Table 13: Critical Values of Spearman's Rank Correlation Coefficient*

Glossary

A

Alternative hypothesis The statistical statement of the population parameter values that we believe are true; denoted H_a or H_1.

***A priori* probabilities** Those probabilities determined by using theory or intuitive judgment as opposed to experimentation.

Arithmetic mean The most widely used measure of central tendency. We obtain the mean, μ, of a population by summing the values of the observations, X_i, and dividing by the number, N.

Array A set of data ordered from the smallest to the largest value.

Average A number used to represent the central value or central tendency of a data set or a distribution.

B

Bernoulli trials Trials that involve random processes for which there are only two possible outcomes that occur without any fixed pattern, and the probability of either outcome remains fixed for each trial.

Bias Applies to an estimator. The bias is zero if the expected value of the estimator equals the population parameter; i.e., if $E(\bar{x}) = \mu$ for the sample mean; otherwise the amount of the bias is given by the expected value minus the parameter.

Binomial pdf The probability density function that comes from Bernoulli trials.

Blocks Account for some systematic source of variability within the data and remove it from the experimental error by placing it within the blocks, such as selecting fields as blocks in a variety trial experiment because they have different soil types.

C

Cluster sampling First select groups of individual items from the population at random, called clusters, and then choose all or a subsample of the items within each cluster to make up the overall sample.

Coefficient of variation Coefficient that determines the relative variability in a set of data. It is the ratio of the standard deviation to the mean, multiplied by 100 percent.

Combinations The number of different arrangements in a group of things that do not depend on order; each arrangement contains a different group of the items.

Complement It means *not;* thus *A* complement is not *A*, but all of the other events in the set.

Composite index Combines information from several sets of data into one index.

Compound event An event composed of several mutually exclusive events.

Conditional probabilities Probabilities associated with events in a subpopulation.

Confidence interval Probability-based interval estimates of a population parameter, such as μ, constructed from the sampling distribution of an estimator such as $\overline{X}$.

Consistent estimator $\hat{\eta}$ concentrates completely on its target as sample size increases indefinitely toward infinity; i.e., it becomes unbiased and its variance approaches zero. As sample size becomes infinite, a consistent estimator $\hat{\eta}$ provides a perfect point estimate of the target η.

Continuous variables The result of measurement; hence they can assume any value in the range of the variable, depending upon the accuracy of the measuring instrument.

Convenience sampling A sample selected based on our convenience and without regard to whether the population is adequately represented.

Correlation Measure of degree of association between two random variables, *X* and *Y*. They are correlated depending upon whether they move in the same or opposite direction.

Correlation coefficient, *r* Measures the strength of the degree of association between two random variables. It ranges in value from ± 1, i.e., $-1 \leq r \leq 1$.

Critical value Table value of the test statistic (Z, t, etc.) that cuts off α (or $\alpha/2$) in the tail of the sampling distribution of the null hypothesis. If the computed value of the test statistic equals or exceeds the critical value, we reject the null hypothesis.

Cross section A set of statistical data that is collected at a single point time over a wide area.

Cyclical fluctuations Movements in the business cycle that reflect the expansion of the economy from a low recessionary period through an economic boom and then its contraction toward another recession.

Degrees of freedom The number of observations in a sample that are free to vary and, hence, take on any value.

Descriptive statistics Numerical data.

Discrete variables Variables that take on whole number values; such variables are the result of counting.

Disjoint Two sets that have no common elements; their intersection is zero.

Dispersion A measure of the representativeness or validity of an average; the amount by which data in a distribution are dispersed away from the average or clustered around it. The smaller the dispersion relative to the average it accompanies, the more representative the average, and thus the more comfortable we are in using the average as the one number to represent the entire distribution.

E

Efficiency A sample design is more efficient than another if it results in lower costs but the same degree of reliability. Also, efficiency refers to the variability of an estimator. One estimator is more efficient than another if it has lower variance.

Equiprobable events If there is no reason to favor a particular outcome of an experiment, then we should consider all outcomes as equally likely.

Estimates Specific values of random variables.

Estimators The random variables from a sample that we select to estimate population parameters, such as $\overline{X}$ to estimate μ.

Event The name we give to each outcome of an experiment that can occur on a single trial. Also, any subset of a universal set of an experiment.

Expected value or mathematical expectation The probability-weighted arithmetic mean of a variable; i.e., the weights are all of the probabilities associated with the values that the variable can assume.

Experiment Any process of observation or obtaining data.

Experimental unit The entity that receives a treatment.

Explained variation The percent of the total variation in the random variable Y accounted for by the set of Xs.

F

Finite population correction factor The factor used with relatively large samples from small populations when computing the standard error of the mean; it is the square root of the term $N - n$ divided by $N - 1$.

Frequency distribution Data summarized with a given number of classes in which each of the classes have a certain width or number of units expressed as an interval. The frequency column in the table expresses the number of observations that fall within each class. The total number of frequencies must equal the number of data points.

Frequency polygon A line graph of a frequency distribution that shows frequencies on the y axis and class midpoints on the x axis. To make the graph touch the x axis, we plot midpoints of imaginary classes at each end of the distribution against zero frequencies. If we leave these out, the graph "floats" above the axis and looks strange.

H

Histogram A bar chart of a frequency distribution; the bars touch and the real class limits appear on the x axis at the end of each bar, and the frequencies appear on the y axis. If the class limits are not all the same width, neither are the bars of the histogram.

I

Intersection The area common to two or more sets or the elements common to those sets.

J

Judgment sampling Use of judgment rather than some objective method such as random numbers for selecting elements of the population for the sample.

M

Mean square A term for a variance commonly used in the analysis of variance.

Median A place average that is not affected by the values of observations at the extremes of the data that is therefore a better average to use in cases in which there are likely to be extreme values such as with income and education data. For ungrouped data, the median is the value of the middle observation when the data are arrayed.

Method of least squares Method that focuses on the random variable Y in regression analysis and minimizes the sum of squared deviations in the Y direction about the regression line; used to obtain estimates of the regression parameters a and b, the intercept and slope coefficients for the equation for the line.

Midrange An average that is the arithmetic mean of the smallest and the largest item in a data set. When we array data, it is the average of the first and last items. It is commonly used in cases in which only the extreme values of a data set are kept, such as average daily temperature and average daily stock prices.

Mode An average that identifies the most common observation in the data set. For ungrouped data, we simply examine the data and select the value of the observation that occurs most often.

Modified mean Obtained by striking out the lowest and highest values in each column and averaging the remaining values; used in calculating seasonal variation.

Multicollinearity When the independent variables in the linear model are highly intercorrelated.

Multiple linear regression Equation that explains the variation in Y with a set of X variables (more than one).

Mutually exclusive The occurrence of one event precludes that of any other event; for example, a head on a single toss of a coin precludes a tail on the same toss, making heads and tails mutually exclusive.

Nominal data Objects are named but not much else. Thus, we can have livestock sorted into calves, heifers, steers, cows, and bulls, etc.

Nonparametric tests Tests in which we are unwilling to make assumptions about the underlying form of the sampling distribution of the statistic we are testing.

Normal equations The equations resulting from application of the method of least squares. One normal equation exists for each regression parameter and the set of equations may be solved simultaneously to obtain the estimated values of the parameters.

Normal *pdf* A bell-shaped continuous *pdf* for the random variable X that can assume any value in the range of plus and minus infinity; it is defined by its mean and standard deviation.

Null hypothesis The statistical statement of no difference, or no effect, or nothing; denoted H_o.

Observations Another term for data. Individual data elements are referred to as observations.

Ogive An S-shaped cumulative frequency curve with cumulative frequencies on the y axis and the upper-class limits on the x axis. Frequencies can be cumulated either on a "less-than" or a "more-than" basis. Ogives provide an easy way to divide the frequency distribution into an equal number of parts such as percentiles, deciles, quartiles, and other quantiles.

Open-ended classes Classes in a frequency distribution that have no fixed upper limit, such as 335 and over, more than 350, etc.; used when there a few large observations in the data scattered over a wide interval that would require too many classes if grouped in regular classes with fixed limits.

Ordinal data Data that express a relationship of order so that ranks may be used. Thus, consumers may rank their favorite foods, etc.

Outcomes The results of an experiment, such as the possible number of dots that can turn up in an experiment of tossing two dice (2, 3, 4, ..., 12).

Paired data Data that are not independent because some factor is present that makes them related; for example, data collected for variety trials on different farms are paired since farmers' management skills, field conditions, etc. are different from farm to farm, and thus affect variety yields.

Parameters Population variables.

Permutation Any ordered sequence of a group or set of things.

Platykurtic distribution Distribution that is flat across the top rather than peaked in shape.

Point estimate One value of an estimator; for example, $\overline{X} = 2$.

Poisson *pdf* A discrete *pdf* defined for an average number of occurrences per unit of time or space and indicating the probability for X occurrences in the next unit of time or space; it is always skewed right.

Population All of the items in the group that we are studying.

Power of the test The probability that the null hypothesis will be rejected when it is false; equals 1 minus β, the probability of a Type II error. The investigator wants the power of the test to approach 1 for values of the alternative hypothesis that he or she is concerned about.

Probability The ratio of the number of events favorable to some event A to the number of events in the experiment.

Probability-based sample A sampling procedure in which every element in the population has a known chance of being included in the sample.

Probability density function (*pdf*) A discrete *pdf* assigns a probability to each possible outcome of the discrete random variable. A continuous *pdf* is a set function that expresses a distribution in which a probability is assigned to a range of values from a continuous random variable.

Probability distribution A listing of the possible values of X and their associated probabilities.

Qualitative variables Variables that refer to some attribute that cannot be measured numerically. Examples include marital status, good or defective parts, kinds of fruit, age group belonged to, etc. These types of variables usually have values arbitrarily assigned to them, such as let $X = 1$ if the ball is black, zero otherwise.

Quantitative variables Objects that can be measured or counted.

Quartile deviation A measure of dispersion used only with the median; indicates the dispersion of the data in the middle half of the distribution.

Quota sampling A sample designed to look like the population with regard to some major attribute. We draw elements of the population until we reach the quota for that aspect of the attribute, and then we choose elements containing a different aspect, and so on.

R

Random sample A sample in which every element in the population available for sampling has an equal probability of being selected.

Random variable Any object whose outcomes are due to chance or are stochastic.

Range A measure of dispersion defined as the difference between the beginning and the end of a distribution when the data are arrayed.

Rank correlation The ordinary correlation computed using ranks of the data and not the original observations.

Real classes Classes in a frequency distribution constructed so that the upper limit of the first class is repeated as the lower limit of the second class, and so on. The classes are continuous for the range of the data and must be used in calculations.

Relative frequency The number of times a certain event occurs in n trials of an experiment.

Residual The difference between the observed value of Y and the estimated value of Y from the regression line; denoted e_i.

S

Sample A subset of the population.

Sample space The set of sample points in which each point corresponds to one and only one possible event; all of the sample points are contained in the set.

Sampling distribution The probability distribution of an estimator or statistic such as the mean. It is obtained by taking all possible samples of size n from the population, computing the estimator (mean) for each, and determining the probabilities associated with each value of the estimator.

Sampling error The difference between the value of the statistic or sample variable and the value of the population parameter or population variable.

Scatter diagram A plot of points of the values of X and Y; used to assess correlation between the variables, patterns in residuals from regression analysis, etc.

Seasonal variations Short-term cycles within the data that complete their duration within the year and recur annually.

Secular trend The smooth upward or downward movement that characterizes a time series over a long period of time, such as 10 years or more.

Set A well-defined collection of things, either real or imaginary.

Simple index Index that expresses the relationship between two numbers with one of them used as a base.

Skewness The relative symmetry of a distribution. If the distribution is perfectly symmetric in shape, then it is not skewed. If the distribution has a tail on the right, it is skewed positively or to the right, and if it has a tail on the left, it is skewed negatively or to the left.

Standard deviation A measure of dispersion used with the arithmetic mean. Its value is based on all of the observations in the data set; it is computed by taking the square root of the arithmetic mean of the squares of the deviations from the mean.

Standard error of the estimate A numerical measure of the dispersion of the points around the regression equation; it is computed as the square root of the mean square error from the analysis of variance for the regression line.

Standard error of the mean The standard deviation of the sampling distribution of the means.

Standard normal curve A particular normal distribution with mean 0 and standard deviation 1; probabilities for it are tabled and all other normal curves can be converted to it by means of the Z formula.

Stated classes Classes in a frequency distribution constructed so that there is a break of one unit between the upper limit of the first class and the lower limit of the second; e.g., $215 - 234$, $235 - 254$, etc.; sometimes used in presentations because they cause less confusion in interpretation.

Statistical hypothesis An assumption we make about some parameter or population variable.

Statistical inference The development and application of procedures and techniques for collecting, analyzing, and interpreting numerical data so that the reliability of the conclusions drawn from this process may be evaluated objectively with probability statements.

Statistics Sample variables.

Stratified sampling Sampling in which we know something in advance about the population and use it to partition it into various layers or strata. We choose the strata so they contain similar characteristics and are more homogeneous than the population as a whole. We sample each stratum using simple random sampling and form the overall sample with subsamples from each stratum.

Sturges's rule A formula that solves for the number of classes, K, to use when constructing a frequency distribution; $K = 1 + 3.322(\log_{10} n)$, where n represents the total number of frequencies. The value, K, will usually not be a whole number, so we need to choose whether to round the number up or round down.

Sufficiency The estimator $\hat{\eta}$ uses all the information about the population parameter η that is contained in the sample data. Thus, in calculating the sample mean, all of the values in the sample are used.

Sum of squares The numerator of the formula for the variance; the sum of squared deviations from the arithmetic mean.

Systematic sample We have access to a list of the population and obtain the sample by taking every kth unit in the population, where k stands for some whole number that is approximately the sampling ratio N/n.

Time series A set of statistical data that is collected, recorded, or observed over successive periods of time.

Treatment Any random variable that the researcher controls.

Type I error Error in which we reject the null hypothesis when it is true; done with probability α.

Type II error Error in which we accept the null hypothesis when it is false; done with probability β.

U

Union The elements contained in one set or a second set or those common to both sets.

Universal set All of the events that comprise an experiment.

Variable Any object that can take on a range of values.

Variance The square of the standard deviation, also called a mean square in the analysis of variance. We must use the variance in mathematical calculations, and then at the end, compute the standard deviation if that is what we want.

Variation within samples A term commonly used in the one-way analysis of variance to refer to the error sum of squares or the error mean square.

Weighted arithmetic mean Calculated using weights on each observation other than one. The mean of a frequency distribution uses frequencies to weight each class midpoint, for example.

Index